JAAWS is abstracted or indexed in *PsycINFO/Psychological Abstracts, CAB ABSTRACTS, CAB HEALTH,* and *Wildlife Review Abstracts.*

First published 2001 by Lawrence Erlbaum Associates, Inc.

Published 2018 by Routledge
2 Park Square, Milton Park, Abingdon, Oxon OX14 4RN
52 Vanderbilt Avenue, New York, NY 10017

Routledge is an imprint of the Taylor & Francis Group, an informa business

ISSN 1088–8705
ISBN 13: 978-0-8058-9690-9 (pbk)
ISBN 13: 978-1-138-46265-6 (hbk)

JOURNAL OF APPLIED ANIMAL WELFARE SCIENCE

VOLUME 4, NUMBER 3, 2001

SPECIAL ISSUE:
Food Animal Husbandry and the New Millennium

GUEST EDITORS:
James A. Serpell and Thomas D. Parsons

JOURNAL OF APPLIED ANIMAL WELFARE SCIENCE, *4*(3), 169–173

INTRODUCTION

Food Animal Husbandry and the New Millennium

James A. Serpell and Thomas D. Parsons

School of Veterinary Medicine
University of Pennsylvania

The four articles comprising this special issue of the *Journal of Applied Animal Welfare Science* (*JAAWS*) originally were presented at the conference "Food Animal Husbandry and the New Millennium: Ethical, Environmental, and Societal Impacts," organized jointly by the Center for the Interaction of Animals and Society, and the Center for Animal Health and Productivity at the University of Pennsylvania. Two of the articles explore specific animal welfare concerns associated with modern farming methods; two address the theme of changing public perceptions of animal agriculture in general. Together, they provide an illuminating and critical appraisal of the food animal industry at the dawn of the 21st century.

The articles by Duncan and Rushen focus, respectively, on animal welfare challenges in the poultry and cattle industries. Duncan is unequivocal in attributing the exceptionally high incidence of welfare problems in poultry to agricultural intensification. His review of the most serious of these problems—starvation of laying hens to expedite molting and the resumption of laying, degenerative joint disease in fast-growing broilers, the poor welfare conditions of poultry immediately prior to slaughter, and the particular plight of spent hens whose bones fracture due to osteoporosis—pulls no punches.

Duncan also is uncompromising about where to point the finger of blame. Economic pressures to produce inexpensive meat products have resulted in crowded housing systems; the use of husbandry practices such as food restriction; little incentive to protect the health and welfare of low-value animals at the end of their productive lives; and intensive genetic selection for enhanced production traits without sufficient regard for some of the phenotypic costs, including abnormal behavior. Although critical of this system, Duncan's stance is neither antiagriculture nor provegetarian. He accepts that some problems are intractable, but he also maintains that substantial improvements are possible if the industry is willing to recognize and correct deficiencies and if consumers, in some cases, are prepared to pay more for meat and eggs produced by more welfare-friendly systems. This will require the development of accepted standards of animal care and laws and regulators to uphold these standards. In support of his argument, however, he points to Europe, where such developments are already well underway—for example, the gradual phasing out of battery cages for laying hens and of stall housing for gestating sows.

Rushen, overall, is more guarded in his critique of intensive farming methods in the cattle industry, primarily because the scientific evidence is less clear-cut. He chooses to focus on the prevalence of mastitis and lameness in dairy cattle, because these two health problems are thought to be very painful, are certainly detrimental to productivity, and are increasingly common among modern dairy herds. Despite the evident shortcomings of the data, Rushen nevertheless points to mounting evidence of a correlation between genetic selection for high milk production and the incidence of both maladies, as well as an interaction with modern housing conditions. To tackle these problems, he advocates increasing the emphasis on genetic selection for health-related traits rather than milk yields and decreasing the reliance on high-yielding breeds. Like the welfare-friendly farming methods proposed by Duncan for poultry production, Rushen's suggestions for dairy production promise to increase the cost of milk. Regarding the relation between welfare problems and housing conditions, his recommendations are less definitive, although he suggests that concrete flooring and lack of access to pasture seem to exacerbate both conditions. If proven true, this correlation is an important one, as the majority of the dairy farms in the United States have moved away from pasture-based rearing to more confined housing systems.

The contributions by Fraser and Thompson offer contrasting, yet complementary, views of the gradual deterioration of public trust and confidence in animal agriculture over the last 30 years. Although perceived declines in the welfare of food animals have been an important ingredient of this loss of faith, especially in Europe, such ethical concerns also have become linked inextricably with a host of other, more human-centered fears. These include real and apparent threats to public health (food safety, dietary fat consumption, infectious disease outbreaks, and nontherapeutic use of antibiotics and hormones to enhance animal productivity); damage to local and global environments (pollution of rivers and estuaries, production of greenhouse gases, soil erosion, and tropical deforestation); and deleterious social effects on the structure, stability, and quality of life of rural communities. Fraser and Thompson frame their arguments very differently. Both agree, however, that the root of the public's loss of confidence in agriculture lies in the postwar trend from relatively small, family-owned and -operated farms toward increasingly large-scale intensive or confinement production systems in which many farmers work under contract for a restricted number of large or very large agribusiness corporations.

To Fraser, this historically rapid process of agricultural industrialization has produced a disquieting mismatch between popular ideas or mythologies about what farming ought to consist of—involving long-established pastoral and agrarian ideals of responsible stewardship—and what it actually has become: an increasingly technological, corporate enterprise driven almost entirely by market forces. Therefore, the resulting public pressure for a return to more traditional husbandry methods is, to some extent, an attempt to reduce the dissonance between these disparate, and ultimately incompatible, visions of animal agriculture. Fraser accepts that intensive farming arose in response to the postwar demand for more and cheaper food, and that it has created substantial economies of scale and improvements in animal disease control, nutrition, and productivity. He also stresses that these technological developments, originating as they did in the context of 1950s attitudes to animals, are no longer culturally appropriate. Food animal agriculture, despite being efficient and profitable, is now increasingly out of step with prevailing societal values and concerns and is in grave danger of losing the favorable public image it hitherto has enjoyed. The current, highly polemicized propaganda battle between proponents and opponents of modern farming only makes matters worse by bombarding already bewildered consumers with flatly contradictory messages. The best solution to this impasse, according to Fraser, is for governments to build consensus and develop new policies based on the findings of high-quality, publicly funded research on the actual, as opposed to the perceived, effects of modern animal production on consumers, producers, animals, and the environment.

Thompson reaches conclusions similar to Fraser's, but he arrives there by a different route. He approaches the perceived welfare problems of agricultural animals by viewing them as one element within a "nexus of livestock production risks" that affect everybody with an interest in the outcomes of intensive farming. These risks can be conceptualized in different ways—some more constructive than others. Thompson is unenthusiastic about what he calls "the industrial paradigm," according to which all of the problems associated with modern agriculture, including threats to animal welfare, are unintended but inevitable by-products of the process of industrialization. This way of framing the problem, he argues, leads people to view industrialization as a kind of disease with symptoms, such as animal cruelty, that need to be treated or cured. It encourages people to hark back to the supposedly safer, less abusive, more "natural" production methods that existed before intensification. In Thompson's view, this way of characterizing the problems of animal agriculture is inappropriate not only because it is oversimplified and inaccurate, but also because it leads to political polarization and mutual distrust between those on opposite sides of the debate. Instead, he favors a "postindustrial paradigm" in which social problems such as intensive farming are analyzed in terms of the various risks affecting the concerned parties (including producers), and which then are weighed and evaluated openly to engender feelings of mutual trust and cooperation. Such analyses also may take into account cultural values—such as consumer concerns about welfare of food animals—that are not amenable to scientific measurement.

Fraser's and Thompson's call for open, political consensus building on the real benefits and costs of intensive farming appears particularly apposite in the wake of the public scrutiny afforded the British livestock industry during its recent Bovine Spongiform Encephalopathy (BSE) and foot-and-mouth disease epidemics. Irrespective of the validity of claims that intensive farming practices contributed to the origin and spread of these diseases, it is apparent that such arguments resonate strongly with a public already disillusioned with numerous other aspects of agricultural production. From the perspective of agricultural interests, these epidemics are seen primarily as economic disasters that threaten the survival of large sectors of the industry. To the average man or woman in the street, however, the untimely slaughter of millions of innocent animals on the farm also carries the message that something is profoundly wrong with the way in which modern agriculture is practiced—so wrong, some would argue, that a complete transformation of the entire system is needed. Valid concerns about the psychological well-being of the animals maintained within this system only serve to reinforce these jaded perceptions.

In the absence of economic incentives, one can question the ability of corporate farming interests to take a long-term, sustainable view that would include substantial improvements in the welfare of food animals. Nevertheless, growing public anxiety about how its food is raised or produced promises eventually to force the agricultural industry to address these consumer concerns. As Duncan points out, North America tends to lag behind Europe by about 8–10 years in the field of animal welfare, and significant legislative reforms to protect the welfare of poultry and swine are already either in place or in the pipeline within the European Union. Thus, as the 21st century begins, American food animal industries find themselves at a crossroads. Continued inaction or reactive behavior toward animal welfare concerns ultimately may pave the way for potentially economically crippling legislative initiatives. Alternatively, a proactive stance on animal welfare issues provides the livestock producer with the opportunity both to develop and market animal welfare-friendly products, and to engage in a meaningful dialog on the reforms that may come to dictate both their lives and their livelihoods.

The articles in this special issue of *JAAWS* highlight pressing animal welfare challenges found in several food animal industries and propose conceptual approaches around which to initiate a constructive discussion on food animal husbandry and the new millennium.

ACKNOWLEDGMENTS

We thank the Provost's Interdisciplinary Seminar Fund of the University of Pennsylvania for generously supporting the conference from which these proceedings are derived.

JOURNAL OF APPLIED ANIMAL WELFARE SCIENCE, 4(3), 175–190

ARTICLES

Farm Animal Production: Changing Agriculture in a Changing Culture

David Fraser

Agricultural Sciences and Centre for Applied Ethics
University of British Columbia

Western attitudes toward animals have undergone a gradual evolution during recent centuries, driven by the scientific recognition that humans and many other species share a common anatomical template, a common phylogenetic ancestry, and certain similarities in their social and emotional lives. This evolving view has been accompanied by a heightened popular respect for animals, which has caused increasing opposition to the relatively utilitarian treatment of animals in modern farming. Western culture also tends to venerate the pastoralist image of humans caring diligently for animals, and North Americans tend to venerate the agrarian life-style of farm families living in harmony with domestic animals and the land. These positive images, which have traditionally lent legitimacy to animal agriculture, have been diminished by changes in production methods during recent decades. The resulting debate between critics and defenders of modern animal production has led to widespread confusion and concern about how animal agriculture affects animal welfare, human health, the environment, and world food security. To resolve this situation will require research to create an accurate understanding of the diverse effects of modern animal agriculture, together with measures to harmonize agricultural practices with changing public values.

According to the legends of the Ojibwa people of central North America, humans were able to inhabit the world because of the cooperation of animals, beginning with the turtle whose shell formed the base of the land, and the toad who

Requests for reprints should be sent to David Fraser, Animal Welfare Program, Faculty of Agricultural Sciences, University of British Columbia, 2357 Main Hall, Vancouver, Canada V6T 1Z4. E-mail: fraserd@interchange.ubc.ca carried soil up from the depths. For the people of ancient Egypt, human destiny after death was decided by the god Anubis (who was part jackal), the goddess Matte (who was part falcon), and other deities who spanned the human, animal, and supernatural realms. In the opening narratives of the Biblical book of Genesis, all beings were created by a single and omnipotent God who then gave humans the responsibility of ruling over the other species.

As these examples illustrate, there is a widespread tendency for cultures to possess an "animal mythology" that helps to define the relationship between humans and other species. Animal mythology, in this sense, is not a pejorative term implying incorrect or antiquated ideas. Rather, it refers to fundamental popular beliefs and values regarding animals, often embedded in a culture's art and stories, which influence how people view animals and what they judge to be proper conduct toward them. This mythological view is roughly the opposite of the reductionist view often encountered in science. The reductionist view sees animals as composed of constituent parts, such as genes and physiological processes. The mythological view takes the animal as a whole and then adds further attributes, including values and symbolic meaning, which resonate with human fears and aspirations.

Today in commercial agriculture and the food industry, animals are often viewed as commodities to be produced, processed, and traded. However, this view of animals clashes, to a degree, with the animal mythology of our culture, and that dissonance leads to conflict and moral concern. Thus, to understand society's current disquiet over the use of animals in agriculture, we need first to understand the underlying animal mythology on which our values are based. However, what are the animal mythologies that pertain to farm animals in our culture? How are these mythological views evolving? Are there ways to reconcile our use of agricultural animals with our animal mythology?

CHANGING VIEWS OF ANIMALS IN WESTERN MYTHOLOGY

Fundamental to any animal mythology are beliefs about the nature of animals and correlated beliefs about their value and moral significance.

In the West, the most highly valued animal has traditionally been the domestic dog which is consistently depicted in art and literature as a loyal, sympathetic, and intelligent helper to humankind. Consistent with this view, dogs are often treated as members of human families, given distinctive names, rescued from abuse by public institutions, and totally exempted from slaughter for human food.

Ironically, the lowest end of our value scale has traditionally been occupied by the dog's close relative, the wolf. Through much of Western history, the wolf has been portrayed as an arch-enemy of the human race who, in numerous folk tales, connives to eat children and old people, and whose death, however gruesome, is invariably a source of satisfaction. In line with this negative image, people for centuries have hunted, trapped and poisoned wolves with few scruples, and until recently public funds in North America were spent to encourage the wolf's extermination.

The farm animals fall between these two extremes. They are seen as a part of human culture and sometimes a source of pride, although nonetheless being valued mainly for their practical usefulness. Hence, they are seen as worthy of care, but generally in ways that are consistent with their utility.

Within our value system, therefore, it might seem completely appropriate for a person to take an aging dog to a veterinarian to prolong the animal's life, then carefully (so as to avoid causing undue stress) load a truck with adolescent pigs to be shipped for slaughter, and finally set out a leg-hold trap to do away with some pesky coyote. Objectively, those three species are roughly similar in their level of mental functioning and their capacity for suffering—attributes that we might view as providing the logical basis for considering animals worthy of moral concern. The fact that we treat the species so differently shows the pervasive influence of our animal mythology.

Nonetheless, even deeply rooted traditional beliefs such as these are open to change. As one example, Dunlap (1988) describes the radical revision of our mythic understanding of the wolf. As Dunlap notes, increased study of wolves—their behavior, communication, and role in ecological systems—led to the wolf shedding its image of treacherous villain, and coming to be seen instead as an intelligent, family-living animal serving vital ecological functions. In line with this about-face, public funding for the killing of wolves was withdrawn, and now public funds are even being spent to preserve wolves and to reestablish them in parts of their former range.

Whales have benefitted from a similar mythic make-over. Traditionally, whales were little more than shadowy bit-players in Western culture, appearing only occasionally in folk-tales as mysterious monsters or dangerous adversaries, and in practical terms treated largely as mobile blubber factories. In recent years, however, we have seen a greater understanding of whales' ability to migrate over vast distances, their complex communication, their social and cooperative living, and their care of their young. Patchy as this understanding remains, it has contributed to a radical change in the value we attach to whales, to the point that confining whales, even for entertainment and education, has become a cause of intense controversy. Ironically, we are even seeing North Americans of European descent protesting against the killing of whales by North Americans of aboriginal descent, suggesting that the whale has become, in effect, a sacred animal for certain European Americans.

THE ROLE OF SCIENCE IN OUR EVOLVING MYTHOLOGY

This revaluing of the wolf and whale are examples of a widespread process that has been underway for several centuries, involving changes in our empirical beliefs about the nature of animals and correlated changes in our ethical beliefs about how animals should be treated. One of the longest-running debates in Western thought—dating back to ancient Greece, and reemerging repeatedly over the millennia—is whether humans are unique and set apart from all other species, or whether we are simply one species among many, with close links to the animal world (Harwood, 1928; Preece, 1999; Sorabji, 1993). At one end of the debate were influential figures such as Descartes and Kant, who saw humans as unique, saw animals as having been put on earth purely for human use, and therefore considered that we have only minimal responsibilities toward them. Opposing this view were other influential figures such as Voltaire and Goethe, who emphasized our relatedness to the rest of nature, and recoiled from the idea of treating animals as if they were purely for human use.

A few centuries ago, those who emphasized human uniqueness could draw on several widely shared beliefs about the nature of animals. One involved appearance: Animals had four legs and fur, or wings and feathers, or fins and scales, and were not at all like the smooth-skinned bipeds who were designed in the image of God. A second involved historical origin: The Biblical tradition held that all other species resulted from a separate act of creation, with humans fashioned either before (Genesis 2) or after (Genesis 1) nonhuman animals, and living in a special relationship with the Creator. A third involved mental and spiritual life: Animals were often viewed as intellectually inferior, as not using rational thought, and even (according to some commentators) as having physical bodies but no souls (Harwood, 1928; Preece, 1999).

Over the centuries, science has slowly eroded these claims to human uniqueness, starting with the notion of unique appearance. With the rebirth of learning in Europe, comparative anatomy became one of the frontiers of science, and through centuries of anatomical research, the homologies of the vertebrate body came to be recognized. Moreover, dissection was a form of educational entertainment. Beginning in the 13th century, dissecting "theaters" sprouted up in most of the major European centers of learning, and these attracted spectators who would pay to attend the dissection of an animal or, better yet, of a human criminal cut down fresh from the gallows. As a result, the growing anatomical knowledge worked its way into popular understanding and by roughly 1700, according to historian Dix Harwood (1928), the fundamental anatomical similarity between humans and other species had become widely recognized.

A second barrier fell with the development of evolutionary thinking during the 1800s. Through Charles Darwin and his contemporaries, people began to see the human species sharing not only a common anatomical template with other species, but a common ancestry as well. This idea, although resisted by many who saw it as a direct challenge to the special status of humans, has gradually become the dominant view in society and has influenced what we see as appropriate treatment of animals (Rachels, 1990).

During the late 1900s, the study of animal behavior has led to a further revision of our view of animals, this one centered on mental and emotional experience. Seminal figures in this development were primatologists such as Jane Goodall (1971), Dian Fossey (1983), and Biruté Galdikas (c1995). In their research, done more in the manner of cultural anthropology than of ethology, animals were viewed not simply as exemplars of their species, serving as data points to estimate the normal or average, but more as persons possessing individuality, unique life histories, and complex social and emotional lives. For example, from Goodall (1971) we learn of McGregor, a chimpanzee who was stricken with polio in adulthood and tried pathetically to reestablish friendly relations with old acquaintances who recoiled from his disability. We also learn of Flint, an eight-year-old who remained so attached to his aging mother that when she died, he stayed near her death place until he, too, died of starvation. These field studies—made accessible to the public by popular books and the media —have been accompanied by other novel types of research. In books such as *Reaching into Thought: The Minds of the Great Apes*—a title unimaginable for a scientific book 30 years ago—we see psychologists borrowing the classic approach of Piaget to study the stages of cognitive development in other species (Russon, Bard, & Taylor Parker, 1996). In *Next of Kin* (Fouts, 1997) we learn about the author's experiences in communicating with chimpanzees by American Sign Language. Summing up these developments, Fouts notes that modern research has led people to see the chimpanzee as, "a highly intelligent, co-operative, and violent primate who nurtures family bonds, adopts orphans, mourns the death of mothers, practices self-medication, struggles for power, and wages war" (p. 58). The gap between humans and other species could hardly seem more narrow.

This altered popular understanding of animals is stimulating a major rethinking of what constitutes ethical conduct toward them. A few decades ago, it was socially acceptable in our society to shoot a chimpanzee mother to capture her infant, raise it in a cage, and use it as a living test-crash dummy. Today, in contrast, we see an international movement to ban all harmful use of the Great Apes in biomedical research (Cavalieri & Singer, 1993). Moreover, heightened concern for prominent animals such as wolves, whales, and chimpanzees is symptomatic of a much broader revision of human thinking about the nature of animals, a revaluing of their worth, and a serious questioning of ways of treating animals that seemed uncontroversial in earlier decades. The implications for animal agriculture are profound.

THE AGRICULTURAL USE OF ANIMALS
IN WESTERN MYTHOLOGY

The agricultural use of animals involves a mythology of its own, including at least two deeply rooted ideas that have had a major effect on people's perception and acceptance of animal agriculture.

A key element is the image of the virtuous pastoralist, inherited in part from the Judeo-Christian Bible. In the early biblical culture, the raising of domestic flocks and herds was an important economic activity; hence, it is not surprising that this culture legitimized the ownership and use of animals. Nonetheless, for pastoralists to prosper, these living possessions had to be given appropriate care: They had to be rested in green pastures, led beside still waters, defended when in danger, even nursed back to health when injured. These practical demands of pastoralist life were reinforced by cultural norms that attached great value to the diligent care of animals. For the biblical David, the first indication that he would become a great king was his care and courage in protecting his father's sheep. The sign that Rebecca had been chosen as the wife of Isaac and mother of her nation was her offer, when asked for water by a thirsty stranger, to water his camels as well. In addition, of course, a diligent shepherd protecting a flock of sheep was such a positive image that it was used as a common metaphor for divine goodness. Thus, the Biblical culture put the conscientious pastoralist on a moral pedestal, and the raising and killing of animals was seen as a legitimate—even virtuous—activity as long as it occurred within that context (Preece & Fraser, 2000).

A second element of our agricultural mythology is a widespread tendency, particularly in North America, to venerate farm families living in a harmonious relationship with the land. As philosopher Paul Thompson (1998) notes, this "agrarian ideal" is deeply rooted in American political philosophy. At the time of American Independence and the contemporary political upheavals in Europe, there was serious debate as to whether democracy was a workable form of government. Critics claimed that democracy was bound to fail because the common people would vote themselves benefits that the nation could not support. On this basis, countries like the United Kingdom tried to balance the power of the elected parliament with a second tier of government consisting of hereditary land owners whose connection to the land was thought to give them a commitment to the future of the nation. In response to this concern, Thomas Jefferson claimed that democracy is feasible in the New World because many ordinary citizens would own land and would therefore exercise their democratic powers judiciously. Thus, as Thompson notes, American political thought linked agrarian living with the ideals of democracy and citizenship.

Unlike pastoralism, agrarianism was not defined principally in terms of human care of animals, but animals were still an important part of agrarian life. To begin with, Thompson (1998) notes, farm animals were an integral part of the ecology and economy of the farm, with the different species serving important and complementary roles. Animals were also important for moral education because children often learned responsibility by caring for animals. In addition, animals on traditional farms were seen as living natural and wholesome lives, much as the human members of the agrarian family were seen as living natural and wholesome lives removed from the artificiality of the city. Thus, the agrarian ideal puts the family farm on a moral pedestal, and again the raising and killing of animals is seen as a legitimate or even virtuous activity as long as it happens within that context.

In summary, our culture has at least two deeply rooted beliefs about the relationship between people and farm animals: a pastoralist ideal that venerates diligent animal care and an agrarian ideal that reveres the farm family living a wholesome life in harmony with the land and their animals. Inasmuch as modern animal production is perceived by the public as conforming to one or both of these ideals, it is almost guaranteed a certain level of public trust and approval. This popular mythology has helped set the stage for a vigorous battle to mold public perceptions of modern animal agriculture.

THE REVOLUTION IN ANIMAL AGRICULTURE

Before turning to the battle, let us briefly survey the disputed territory. Until about 1950, animal agriculture in the industrialized countries involved fairly traditional methods that relied heavily on labor to accomplish routine tasks such as feeding and removal of manure, and generally involved keeping animals in outdoor or semioutdoor environments. After World War II there emerged a new generation of technology typically called "intensive" or "confinement" animal production. Confinement systems generally use hardware and automation instead of labor for many routine tasks, and the animals are generally kept in specialized indoor environments (Fraser, Mench, & Millman, 2001). In most industrialized countries, confinement systems are now the norm for swine and poultry; dairy cattle are generally kept in semiconfinement systems where the animals have some access to the outdoors during at least part of the year, and beef cattle and sheep are generally kept in nonconfinement systems during much or most of the production cycle.

In the most restrictive of the confinement systems, animals spend most or all of their lives indoors with large numbers of others, and their freedom of movement and opportunity to perform natural behavior are greatly limited. For example, laying hens are generally housed in cages containing 3 to 10 birds, often with thousands of cages per barn. The cages allow automated feeding, collection of eggs, and removal of manure, but they typically provide minimal space and no opportunity for the birds to perform such natural behavior as dust-bathing, resting on a perch, or retreating to a secluded nest-box for laying. Pigs are generally housed in group pens, often in totally or partially enclosed buildings with unbedded concrete floors. The pens generally provide just enough space for animals at market weight to lie down simultaneously, and little opportunity for them to exercise or to perform natural behavior such as exploration and rooting. Pregnant sows are typically kept for 2 to 3 months in individual stalls or "gestation crates"; these prevent aggression between sows and make it easy to feed the animals individually, but the stalls provide only enough room for the sow to take about one step forward and back, and not enough to walk or turn around.

While these changes were occurring in animal housing, other performance-enhancing methods also came into widespread use. Vaccines and other disease-prevention measures eliminated or reduced the incidence of certain diseases and led to concomitant increases in productivity. Better diet formulation and control over food intake eliminated certain nutritional problems. Other measures included the use of hormones (as implants, feed additives or injections) to increase growth rate and efficiency, and the use of antibiotics as feed additives to enhance growth. Intense genetic selection also resulted in major increases in certain production traits including the milk yield of cows and the growth rates of pigs and broiler chickens.

From an animal welfare viewpoint, these changes in animal housing and production methods reduced or eliminated certain problems, but exacerbated or created others. Indoor housing has eliminated many of the problems related to predation and cold weather, which were major causes of death in older systems. On the other hand, indoor housing generally exacerbates problems of hot weather, and creates new problems of exposure to harmful levels of dust and gases in the air. Separation of animals from excreta and soil-borne pathogens has helped control certain infectious diseases, but the confinement of so many animals together has potentially increased the opportunity for transmission of others. Genetic selection has in some cases led to enhanced resistance to disease, and in other cases to new health problems; for example, selection for rapid growth in broiler chickens can lead to leg abnormalities and lameness, and selection of laying hens for high egg yield can produce birds that are prone to osteoporosis. Thus, some of the major animal welfare problems of earlier times were reduced or eliminated by the changes in animal agriculture; whereas other animal welfare problems, including those involving comfort, freedom of movement, and opportunity to perform natural behavior, have become much more common.

In the developed countries, the technological changes in animal agriculture have generally been accompanied by changes in farm structure and the sociology of rural communities. For species raised in confinement, there has been a steady increase in the size of farms, and a corresponding decrease in the number of people directly involved in animal production. In the 1950s, for example, the average flock of laying hens in the United States contained fewer than 1,000 birds, whereas flocks of tens of thousands are now common, and flocks of millions are not unknown. For many commodities and regions, these changes have involved mainly an increase in the size of individually or family-owned farms. However, in some regions and sectors, notably for certain commodities in the United States and some of the former Soviet countries, large, corporately or collectively owned units have replaced many traditional family-owned units. An extreme example is the United States broiler chicken industry where only five companies controlled more than half of the market by the end of the 1900s (Fraser et al., 2001).

These changes have not occurred in a uniform way. For example, Europe does not allow hormone treatments to enhance the growth or milk production of cattle. Although very large units have become the norm for egg production in the United States, individually or family-owned units remain the backbone of the industry in Canada. Of the world's two leading pig-producing countries, the largest (China) uses mainly traditional, small-scale production, whereas the second largest (the United States) uses mainly confinement systems. Even within a single sector and country, the industry can be diverse. For example, in California (the top dairy producing state in the U.S.) nearly 80% of cows are kept on farms with more than 500 cows, whereas in Wisconsin (the second highest dairy producing state) only about 5% are on farms of such size (USDA National Agricultural Statistics Service, 2000). Thus, the revolution in animal agriculture has not occurred (as is sometimes claimed) as a unified package, whereby confinement housing, biotechnology, large farms, and corporate ownership necessarily go hand in hand. Rather, we see a constellation of changes differing between sectors and regions, and having varied and far-reaching effects on the commercial raising of animals (Fraser et al., 2001).

CONFLICTING PORTRAYALS OF ANIMAL AGRICULTURE

Given the increased value attached to animals in our society, some of the changes in animal agriculture have provoked a range of reactions including concern, criticism, and in some cases open opposition to the commercial raising of animals. Some of the critics and opponents have been active in trying to reshape public perceptions of animal agriculture along lines captured in the following quotations:

1.   In the United States, as elsewhere, factory farming has become a major commercial enterprise that is threatening the family farm with extinction (Dolan, 1986, p. 67).

2.   The problem is that the behemoths of modern agribusiness seek profit without reference to any ethical sensitivity to the animals in their keeping (Robbins, 1987, p. 97).

3.   Whether they are battery chickens in their cages or pigs in sow stalls, all experience the same mental anguish that would drive many humans to suicide—but factory-farmed animals do not have that option (Penman, 1996, p. 25).

4.   Eating meat has been linked to heart disease, cancer, diabetes, arthritis, and osteoporosis. Animal fat and cholesterol ... are the leading causes of heart attacks and strokes. Other health risks are increased by the chemicals, antibiotics, and hormones found in meat. ... Not eating meat, on the other hand, significantly reduces your risk of illness (Fraser, Zawistowski, Horwitz, & Tukel, 1990, p. 38–39).

5.   [Cattle] are destroying the very biosphere itself, threatening the future stability and viability of entire bioregions of the world. Cattle are among the major environmental threats facing the planet today (Rifkin, 1992, p. 191).

6.   The vast tonnage of food fed to animals to supply the rich countries with their heavily meat-based diet is given at the expense of hungry people around the world (Coats, 1989, p. 140).

In these quotations we see a complex, multisided criticism of animal agriculture that has evolved and expanded over the past 40 years. When public disquiet erupted over modern animal production in Britain in the 1960s, the focus was almost entirely on how animal production technology affected animal welfare (Brambell, 1965; Harrison, 1964). Consequently, scientists could respond to the concerns by trying to determine, for example, whether the welfare of chickens would be improved by access to outdoor runs, and whether sows would be better off in group pens than in individual stalls. Today the animal welfare concerns remain. Now, however, opponents of animal production often use animal welfare as one of several elements in an effort to eliminate all images of pastoralist animal care and agrarian families and create an alternative image of greedy, impersonal corporations displacing farmers from the land, exploiting animals, poisoning consumers, polluting the environment, and taking food away from the hungry (Fraser, 2001).

In response to these negative portrayals, some agricultural organizations have promoted a competing image, depicting animal agriculture as fully reflecting traditional pastoralist and agrarian values, while benefitting from modern knowledge and technology. According to these neo-traditional portrayals, modern farming is thoroughly beneficial for animal welfare. Domestic cattle, for example, are said to "live in the lap of luxury," and confinement housing, instead of causing animals to suffer, is claimed "to protect the health and welfare of the animals." Likewise, industry materials claim that traditional animal care values are firmly in place—that producers "have always recognized their moral obligation to provide humane care for their animals", and are "committed to providing the utmost in humane care." Moreover, animal agriculture is said to benefit the environment because grazing "improves vegetation health and diversity", and livestock "complete the nutrient cycle, returning valuable manure to the land". In a similar manner, industry materials systematically contradict each of the various accusations, countering rhetoric with rhetoric, and matching overgeneralization with overgeneralization (quotes from Animal Industry Foundation, 1988; Herscovici, 1996; National Cattlemen's Beef Association, 1998a,b; National Pork Producers Council, 1998).

The situation has, in effect, produced a propaganda battle whereby the public is presented with two simplistic and emotive portrayals of animal agriculture, flatly contradicting each other on a wide range of issues (Johnson, 1991). Moreover, because animal agriculture is so diverse, both sides can select statistics and examples to make their generalizations sound convincing and to cast doubt on opposing claims. As a result, the widespread public concern about the ethics of animal agriculture is compounded by widespread public confusion about the simple facts of what modern animal production entails. When there is such profound disagreement about the facts, there is very little hope of achieving consensus on what would constitute appropriate actions or policy changes. How, then, should we proceed if we want to achieve greater harmony between animal agriculture and changing public values?

A NEED FOR ANALYSIS

Despite the polemicized way in which the issues have been presented, the debate has raised legitimate questions about how modern animal production affects

- the welfare of animals,
- the welfare of animal producers,
- the health of consumers,
- the sustainability of food production,
- the global environment, and
- world hunger and food security.

Although the effects of modern animal agriculture have often been portrayed as either entirely negative or entirely positive, such complex changes to an already complex industry almost inevitably produce a mixture of positive and negative outcomes. Understanding these outcomes is of great importance, not only for animal agriculture and its acceptance by the public, but for animals, for human health and society, and indeed for the future of the biosphere. The issues cry out, albeit belatedly, for careful analysis to understand the effects, and then for appropriate action to ensure that one of our most important and diverse activities develops in desirable and socially acceptable directions.

Given the importance of the issues, it is dismaying that they are not at the top of the agenda for publicly funded agricultural research. Instead, we see a patchy and inadequate research effort, with some promising work on certain topics such as farm animal welfare, but little or none on others. Why is it that such important issues, although attracting so much attention and so many categorical pronouncements, have generated so little real investigation?

The answer is not a simple lack of resources, as the industrialized countries invest substantially in agricultural research. Traditionally, much agricultural research was seen as a public service activity whereby publicly funded scientists attempted to solve problems for the sizeable fraction of the human population engaged in farming. More recently, there has been a shift toward industrial funding of agricultural research (Hodges, 2000), and an increasing amount of research talent is helping businesses to develop and test new commercial products for use in agriculture. In the wake of this shift, many countries have redirected public funding toward these and similar commercial goals. Worthy as this research may be, it has left the issues identified previously, which are of the utmost importance for public policy and the future of agriculture, largely ignored as topics for serious research. To resolve this situation, the agricultural research community needs to regain its sense of serving the broader public, and scientists, research managers and funding agencies need to view the issues raised in the debate over animal agriculture as high priority areas for research.

In researching these issues, there is also a need to establish a tradition of rigorous investigation and analysis—a tradition that has been absent in some academic contributions to the debate about animal agriculture (Fraser, 2001). As one example, in a widely read essay published in the 1980s, an ethicist argued that we should not raise and kill animals for food, giving the following as one of his arguments:

> Many of the peoples of the world are suffering and dying from protein deficiencies. In the United States during 1968, we fed to livestock ... 20 million tons of plant protein that could have been consumed by humans. Although the livestock provided 2 million tons of protein, the 18 million tons of protein "wasted" by this process would have removed 90 percent of the yearly world protein deficit. (Gruzalski, 1983, p. 255)

The simplicity of the argument is beguiling, but the role of animal agriculture in the global food supply actually raises a number of difficult questions. To what extent does undernutrition in the world really result from a "protein deficit," which could be eliminated by increasing available supplies? Would ending the production of animals in a given region actually lead to more grain being produced for human consumption? Under what conditions would on-going food aid lead to long-term reduction of hunger in developing countries, as opposed to undermining local agricultural development?

As a second example, some scientists have long maintained that the pursuit of profit is sufficient to ensure a high degree of animal welfare. For instance, Rosenwald (1981) proposed,

> It is in the best interests of the producer to treat his animals as well as possible to get the greatest economic return and, therefore, there really isn't any basic conflict between the ethics and the economics of poultry production. (p. 578)

Again, the simplicity of the argument is appealing, but the relationship between farm profit and animal welfare actually involves complex questions. Do the levels of amenities (amount of space, frequency of inspection) that maximize profit really maximize animal welfare? Can genetic selection for production traits increase profit but reduce animal welfare? In producing high-value products such as pale veal, are the required animal management practices necessarily in the animal's own interests?

In these two examples, a genuine and knowledgeable analysis of the issues would be an important intellectual and practical contribution. The simplistic statements quoted are symptomatic of a long-standing tendency for philosophers, scientists, and other academics to treat these issues in a superficial rather than an analytical manner. A cultural change is needed within the research community such that scientists and academics stop simply aligning themselves with one side or other in the debate, and instead use their technical, investigative and analytical skills to create a genuine understanding of the diverse effects of animal agriculture.

HARMONIZING ANIMAL AGRICULTURE
WITH OUR ANIMAL MYTHOLOGY

Through our collective experience in international development, we have come to recognize that technology needs to be culturally appropriate—that we cannot simply take a form of technology from one culture and expect it to function smoothly in another. The same insight can help us understand the current conflicts over animal production in our own society. Many of the methods that currently dominate animal agriculture in the industrialized countries originated in earlier decades when social values regarding animals were different from those of today. In the mid-20th century, key concerns for animal agriculture were increased production, increased efficiency, and low food prices. The changes in animal production methods in the last 50 years helped to meet these goals. Today, however, the view of animals in the developed countries has changed greatly, and other issues, including food-related health issues and the environment, have also become more prominent public concerns. There is real potential for animal agriculture to find itself increasingly out of line with public values and hence experience an erosion of public support.

Should agriculturalists be concerned about such a development? Until recently, agriculture in the industrialized countries has enjoyed very favorable treatment by the public, including public funding of agricultural research and extension services, together with other forms of subsidization and support. The industrialized countries have also tended to exempt agriculture from certain types of regulatory control including legal restrictions on methods of animal production. However, if animal agriculture comes to be viewed as operating contrary to widely held public values, then it will lose the favorable treatment it has enjoyed. Thus, the animal industries may work strongly against their own interests if they continue using methods that become increasingly out of line with public values.

How can animal production be harmonized with changing public values? First, industry leaders need to look carefully at changing values, and consider how to respond to the threats and opportunities these create. Are there animal production methods that run strongly contrary to public beliefs about what constitutes appropriate animal care? If so, what kinds of incentives might bring animal production methods more into line? Are there marketing and trade opportunities that might be missed if a country is perceived as having inadequate animal welfare standards? Conversely, are there emerging opportunities to supply products to consumers concerned about animal welfare and other issues? In addition, are there animal protection organizations that will work with animal producers to achieve greater harmonization of production methods and public values?

There is also a need for the industry to support the development of alternative production methods. At present, a few producers and researchers are trying to use and perfect technology such as less restrictive housing for pregnant sows, and noncage systems for laying hens. These efforts tend to receive little industry support, in North America at least, because they depart from the predominant production systems. However, these efforts could potentially pave the way toward harmonizing animal production with public values, and they deserve to be seen as a worthy investment for the future.

Although much could be achieved through far-sighted actions by the animal industries, the challenges posed by changing public values raise important policy issues that will require public involvement and government leadership. We need to decide whether to treat free-market competition as an adequate means of shaping animal production methods, or whether nonmarket measures are needed to protect animal welfare and other key values. We need to decide whether the family-owned farm is of sufficient social and environmental merit that it deserves some degree of protection. Is product labeling a satisfactory means of allowing consumers to influence animal production standards? At what point is regulation appropriate to protect animal welfare, consumer safety, and the environment? The revolution in animal agriculture, despite its far-reaching impact, took place with remarkably little public debate and policy development; and consensus on the issues has failed to emerge. Governments need to create opportunities for the issues to be addressed and studied, and, as much as possible, for consensus to develop on how to reconcile our animal production systems with our changing culture.

ACKNOWLEDGMENTS

This article is based on a lecture at the conference, "Food Animal Husbandry and the New Millennium: Ethical, Environmental, and Societal Impacts," Philadelphia, November 5, 1999, and at the conference, "Farm Animal Welfare Challenge 2000," Guelph, Canada, June 10, 2000. Parts of the article have been developed in more detail in Fraser (2001); Fraser, Mench, and Millman (2001); and Preece and Fraser (2000).

I am grateful to Rod Preece, Joy Mench, and Suzanne Millman for allowing me to borrow from our collaborative work in this text.

This research was supported by the Natural Sciences and Engineering Research Council of Canada.

REFERENCES

Animal Industry Foundation. (1988). *Animal agriculture: Myths and facts*. Arlington, VA: Author.

Brambell, F. W. R., chairman. (1965). *Report of the technical committee to enquire into the welfare of animals kept under intensive livestock husbandry systems*. London: Her Majesty's Stationery Office.

Cavalieri, P., & Singer, P. (Ed.). (1993). *The great ape project*. New York: St. Martin's Press.

Coats, C. D. (1989). *Old MacDonald's factory farm*. New York: Continuum.

Dolan, E. F., Jr. (1986). *Animal rights*. New York: Franklin Watts.

Dunlap, T. R. (1988). *Saving America's wildlife: Ecology and the American mind, 1850–1990*. Princeton, NJ: Princeton University Press.

Fossey, D. (1983). *Gorillas in the mist*. Boston: Houghton Mifflin.

Fouts, R. (1997). *Next of kin*. New York: William Morrow.

Fraser, D. (2001). The "New Perception" of animal agriculture: Legless cows, featherless chickens, and the need for genuine analysis. *Journal of Animal Science, 79*, 634–641.

Fraser, D., Mench, J., & Millman, S. (2001). Farm animals and their welfare in 2000. In: D. J. Salem & A. N. Rowan (Eds.), *The State of the Animals 2001*. (pp. 87–99). Washington, DC: Humane Society Press.

Fraser, L., Zawistowski, S., Horwitz, J., & Tukel, S. (1990). *The animal rights handbook*. Los Angeles: Living Planet Press.

Galdikas, B. M. F. (c1995). *Reflections of Eden: My years with the orangutans of Borneo*. Boston: Little, Brown.

Goodall, J. (1971). *In the shadow of man*. London: Wm Collins.

Gruzalski, B. (1983). The case against raising and killing animals for food. In H. Miller & W. Williams (Eds.), *Ethics and animals* (pp. 251–263). Clifton, NJ: Humana.

Harrison, R. (1964). *Animal machines*. London: Vincent Stuart.

Harwood, D. (1928). *Love for animals and how it developed in Great Britain*. New York: Privately published by the author.

Herscovici, A. (1996). *Food for thought: Facts about food and farming in Canada*. Mississauga, Canada: Ontario Farm Animal Council.

Hodges, J. (2000). Why livestock, ethics and quality of life? In J. Hodges & I. K. Han (Eds.), *Livestock, ethics and quality of life* (pp. 1–26). Wallingford, England: CABI.

Johnson, A. (1991). *Factory farming*. Oxford: Blackwell.

National Cattlemen's Beef Association (1998a). *Animal welfare*. Retrieved December 14, 1998 from the World Wide Web: http://www.beef.org/libref/beefhand/animl.html

National Cattlemen's Beef Association (1998b). *12 Myths and facts about beef production*. Retrieved December 14, 1998 from the World Wide Web: http://www.beef.org/librfacts/myths_facts.html

National Pork Producers Council (1998). *How hogs are raised today—the tradition of care continues on America's hog farms*. Retrieved December 16, 1998 from the World Wide Web: http://www.nppc.org/how.hogs.are.raised.html

Penman, D. (1996). *The price of meat*. London: Victor Gollancz.

Preece, R. (1999). *Animals and nature: cultural myths, cultural realities*. Vancouver, Canada: University of British Columbia Press.

Preece, R, & Fraser, D. (2000). The status of animals in biblical and Christian thought: A study in colliding values. *Society and Animals, 8,* 245–263.

Rachels, J. (1990). *Created from animals: The moral implications of Darwinism.* New York: Oxford University Press.

Rifkin, J. (1992). *Beyond beef: The rise and fall of the cattle culture.* New York: Dutton.

Robbins, J. (1987). *Diet for a new America.* Walpole, NH: Stillpoint.

Rosenwald, A. S. (1981). First European Symposium on Poultry Welfare held. *Poultry Digest, 40,* 576–582.

Russon, A. E., Bard, K. A., & Taylor Parker, S. (Eds.). (1996). *Reaching into thought: The minds of the great apes.* Cambridge, England: Cambridge University Press.

Sorabji, R. (1993). *Animal minds and human morals: the origins of the Western debate.* Ithaca, NY: Cornell University Press.

Thompson, P. B. (1998). *Agricultural ethics—research, teaching, and public policy.* Ames: Iowa State University Press.

USDA National Agricultural Statistics Service. (2000). *U.S. livestock summary, cattle.* Retrieved January 28, 2000 from the World Wide Web: http://usda.mannlib.cornell.edu/reports/nassr/livestock/pct-bb/catl0100.txt

JOURNAL OF APPLIED ANIMAL WELFARE SCIENCE, 4(3), 191–205

Animal Welfare and Livestock Production in a Postindustrial Milieu

Paul B. Thompson

Departments of Philosophy and Animal Science
Purdue University

Structural transformation, food safety, and environmental risks pose challenges to livestock producers. Adjustments to livestock production systems to improve animal welfare will be made in an economic and political milieu characterized by these challenges. However, competing assumptions about contemporary society provide different frameworks for formulating the problems faced by industry and government decision makers. The assumption that industrialization is the key problem in livestock production leads to an application of science that does not adequately address the role of public participation and trust.

The aim of this article is to examine not the welfare of animals on the farm but the social, economic, and environmental parameters under which problems and deficiencies in nonhuman animal welfare must be addressed. If the science of animal welfare is to be applied to livestock, it will be useful to have some general understanding of the circumstances in which livestock are kept. Fraser, Mench, and Millman (2001) provided an overview of the global trend toward industrial livestock production and summarized some of the key welfare issues associated with it.

People who raise livestock for food and fiber must derive a sufficient financial return from their efforts. Thus, the prospects for modifying the care that animals on the farm receive or the environments in which they live are subject to constraints. These constraints are dynamic. Industrial livestock producers in Europe and North America face an array of challenges to their current production, transport, and processing practices. Among them are broad structural transformations

---

Requests for reprints should be sent to Paul B. Thompson, Department of Philosophy, Purdue University, West Lafayette, IN 47907–1360. E-mail: pault@herald.cc.purdue.edu

in the food system, pressure to increase the assurance of safe food supplies, and growing demands for environmental quality (Curtis, 1994; Kunkel, 2000).

Any attempt to address welfare issues must compete with these other problems for resources and for the attention of researchers and policymakers. The following review of these issues is not an apology or defense of the treatment animals receive in existing livestock production systems. It is, instead, an inventory for the conceptualization of what is problematic about contemporary livestock production, and it will not be new to those who are knowledgeable about animal agriculture. The main aim of this article is to compare and contrast two overarching ways to conceptualize the competing issues and challenges that affect livestock production: industrial versus postindustrial society. In contrast to the premises of industrialization, livestock producers are increasingly subjected to an array of forces characteristic of what Beck (1992) called a "risk society." In the postindustrial risk society, both private and public decision making are shaped by a heightened sense of being at risk from the actions of others and a greater willingness for citizens to mobilize politically around issues of risk. The final section of the article illustrates how these two competing ways frame and interpret problems in livestock production—especially including animal welfare.

BROADER CONTEXT OF LIVESTOCK PRODUCTION AND PROCESSING

Contemporary livestock producers and meat processors operate in an economic and political environment that can be characterized by three broad trends: First, production and processing have been undergoing a long-standing structural transformation in the patterns of ownership and control of the primary factors in the production of meat, milk, and other animal products. Second, food safety regulation, which has influenced animal production and processing practices for almost a century, has undergone a period of rapid evolution during the last 2 decades, and more changes may be on the horizon. Third, livestock production recently has been the target of criticism on environmental grounds, and this criticism continues to affect both the regulatory and the political environment for animal agriculture. Each of these three trends is discussed briefly.

Structural Transformation

Any analysis of the broad economic and political environment for livestock production must be set against a backdrop of long-standing trends in ownership and control in the livestock production and processing industries. Any discussion of the way these trends have operated on a global scale would be extraordinarily complex (Wallerstein, 1974). North American structural transformation can be described as a gradual shift away from an ownership structure in which the vast majority of American farms were owner operated, with most of the farm labor coming from family members. Such small, family-run farms have faced increasing pressure for survival in North America (Gebremedhin & Christy, 1996). Before 1900, U.S. and Canadian farmers tended to operate diversified operations producing both crops and animal products. Throughout the 19th century, these were predominantly commercial operations producing commodities for sale on both local and national commodity markets, as opposed to subsistence farms, although many farm families derived a large portion of their diet from their own farm production. Cows, pigs, chickens, sheep, and a goat or a small flock of ducks or geese often were kept mostly outdoors in pens or pastures (though with access to a barn in cold weather). Sales of meat, milk, and eggs each might have contributed a share to family income, although animal products might be just as likely to wind up on the family dinner table (Fite, 1981).

As is well known, animal production is rarely done under such circumstances today. The typical livestock farm is far less diversified, and most livestock producers have specialized in a single commodity. Whereas animals once were kept outdoors in small herds or flocks, many animals now are housed indoors for their entire lives, although both beef and dairy cattle often are kept in open lots. Citing global figures for 1996, Sere and Steinfeld (1996) reported that 79% of poultry, 39% of pork, and 68% of eggs are produced in intensive animal production systems. The primary exceptions to this trend are in sheep and cow–calf production, which have changed far less than other sectors of the livestock industry. Sheep for wool and cattle for beef (and to a much lesser extent for dairy) may continue to be bred and kept on pasture for most of their lives, with beef cattle spending only a few weeks in feedlots for fattening before slaughter. Although beef and dairy cattle and sheep and goats continue to be raised under extensive conditions, industrial practices of early weaning, castration, and dehorning are becoming increasingly common, as is artificial insemination (Fraser et al., 2001). The use of equipment and buildings for confinement and care of animals on the farm simultaneously has increased the economic efficiency of animal production and increased the capital investment required to operate profitably in this sector (Kunkel, 2000).

Contrary to popular misconceptions, most U.S. and Canadian livestock farms continue to be owner operated. However, although families continue to supply most of the labor for many Midwestern grain farms, animal production increasingly depends on the availability of low-wage hired labor, although there is considerable variation in this dependence from region to region and commodity to commodity. Furthermore, although individuals may own their animals, as well as the land and buildings where animals are produced, many of these owner operators are under contract to large corporate entities, especially in the poultry sector. These contracts offer producers relatively little latitude in decision making with respect to the production process, and many who do not produce on contract must follow a very structured feed and husbandry regimen if they are to find a market for their animals. However, these financially small-scale operators typically do bear capital risks associated with the obsolescence of their buildings and equipment. Thus, changes in regulations for animal production—including changes intended to improve welfare—that require replacement or refitting of capital equipment can lead to financial losses for these operators (Kunkel, 2000; Martin, 1997).

Within the processing industry, structural transformations have been underway for many years. The creation of a national railroad network for live transport of animals and refrigerated transport of rendered carcasses led to the development of major centers for assembly and slaughter of animals in Cincinnati, Chicago, and across the Midwest in the latter half of the 19th century (Cronon, 1993). The transport, slaughter, and packing industries have continued to undergo consolidation for decades, although independent packers continued to operate side by side with well-known national firms well into the 1950s and 1960s. The last decades of the 20th century saw the emergence of vertical integration, especially in the poultry sector, with packers supervising and effectively controlling every phase of the production process, down to the specification of and contracting for animal feeds with grain producers. Today, only a few national firms dominate meatpacking in all the major animal commodities. Mergers and acquisitions among these firms now are supervised for antitrust violations (Fink, 1998).

The animal welfare issues associated with livestock transport and slaughter have been the subjects of well-known studies (Grandin, 2000). It is important to recognize that there are also human issues associated with this aspect of animal agriculture. Meatpacking traditionally employed unskilled and low-skill workers and was the site of bloody and contentious unionization activities in the 1920s. After the principal national packing companies signed union contracts, labor relations settled into a period of relative stability through the 1960s. However, during the last 3 decades, labor relations again have become contentious in the packing industry as packers have introduced new technology and work rules, in effect deunionizing the industry. Human rights activists again criticize labor practices in the industry as unsafe, underpaid, and exploitative of ethnic minorities and women. Recent immigrants of Hispanic origin now hold many low-wage packing jobs. The influx of a non–English-speaking workforce has strained schools and government services in Midwestern cities where plants are located and, in some instances, has inflamed racial passions (Fink, 1998).

Food Safety and Nutritional Quality

The safety and quality of meat and milk products were a topic for some of the first consumer-interest advocates. Authors such as Sinclair (1906/1981) raised public consciousness about poor standards in the meatpacking industry, which led eventually to the formation of the Food and Drug Administration and the Food Safety Inspection Service in the U.S. Department of Agriculture. These regulatory actions had little direct impact on animal producers, but they did play a role in early consolidation of the packing industry. Meeting these regulatory standards required modern equipment and trained personnel, giving relatively larger and better capitalized firms an advantage over local butchers and small abattoirs.

Since 1970, however, a number of practices have introduced controversy into animal production at the farm level. Meat consumption itself became the focus of concern as health-conscious consumers attempted to reduce their intake of fat and cholesterol. There has been a shift from red meats to poultry and fish in the United States and Canada. Feed rations have been adjusted, and both breeding and use of genetic markers have been deployed in search of low-fat meat alternatives. Animal producers lobbied to influence government dietary guidelines and discouraged recommendations to lower meat consumption (Levine, 1986). However, dietary changes also have placed producers into sharp, and sometimes acrimonious, competition with one another. Poultry has been touted as preferable to beef, and pork has been advertised as "The Other White Meat." This competition sometimes has made it difficult for producers of different animal commodities to cooperate with one another, even in areas of mutual interest (Kunkel, 2000).

The issues that have been of greatest concern have to do with health risks associated with animal drugs, including antibiotics and hormones, and with food-borne diseases. Livestock producers use pharmaceutical products for both preventative and therapeutic purposes. Although the Food and Drug Administration regulates these uses, they have become somewhat controversial. In particular, producers have been criticized for the nontherapeutic use of antibiotics and of hormones to spur growth rates (Kunkel, 2000; Schell, 1984). Controversy over the use of hormones has been at the root of a long-standing dispute between U.S. producers and the European Union. The approval of recombinant bovine growth hormone in dairy production sparked a heated political debate before U.S. approval, and the product still is not approved for use in Canada or in many parts of Europe. For the most part, scientific support for consumer fears over the use of these products generally is weak and, at best, mixed. Livestock producers and their advocacy organizations generally consider these consumer concerns to be misplaced (Kunkel, 2000).

There is no doubt, however, that a number of serious food-borne diseases have been associated with animal products. Outbreaks of microbial diseases (Salmonella, E. coli O157:H7) have been associated both with failures in the processing and preparation of animal food products and with production practices. Disease outbreaks cause a great deal of economic instability in consumer demand for animal products. Although government guidelines promote that care in the handling and proper cooking of meats is the most effective means for minimizing the risk of microbial diseases, animal producers take precautions to reduce pathogens at all stages of the production process. European problems with new variant transmissible spongiform encephalopathies such as so-called mad cow disease hypothetically are linked to feeding practices introduced in England in the 1970s and 1980s (Brouwer, 1998). Given the array and seriousness of these legitimate food-borne risks, it is not surprising that members of the public are wary of all novel practices, including those that currently appear safe (Thompson, 1997).

Environmental Issues

Animal production has been linked to a number of contentious environmental issues in recent decades, including soil erosion, the production of global greenhouse gases, and the devastation of tropical rainforests (M. A. Fox, 1999). These issues are not trivial, but they relate less to animal welfare than to pollution associated with waste from confined animal production. Animal scientists have undertaken research to address the production of greenhouse gases through adjustment of feed rations and remediation of animal waste (Council for Agricultural Science and Technology, 1992).

The fragility of western rangelands makes many current livestock grazing practices unsustainable throughout much of the western United States. Range scientists calculate that current stocking rates will deplete fragile soils on western rangelands within the lifetime of the current generation of ranchers. In this context, these issues are primarily relevant insofar as they indicate that livestock producers face criticism on environmental grounds wholly apart from the issues of intensive animal production.

Debate over the polluting effects of large-scale intensive animal production is more pertinent in this context, because these are the same industrial production systems that often are the target of criticism by advocates of animal interests. Intensive production systems produce large amounts of animal waste. Nutrients are concentrated in animal waste and are toxic at high levels (Donham, 1998). In extensive production, nutrients from animal waste are cycled back into soils and reincorporated in plant growth. However, nitrogen and phosphorus in animal wastes from industrial production systems often cannot be absorbed readily into soils immediately adjacent to these facilities. The result is a potential for soil and water pollution, and—even under the best of circumstances—animal wastes can produce noxious odors. Possible solutions involve some combining of storage and drying, trucking solid waste away from the facility, and composting. However, these systems are costly to varying degrees and subject to failures that can cause pollution events. Notoriously, flooding can swamp animal waste holding tanks, leading to a release of nutrients. After such events, local water supplies temporarily have contained levels of nitrogen and phosphorous exceeding U.S. Environmental Protection Agency limits (Jackson, 1998; Letson, Gollehon, Breneman, Kascak, & Mose, 1998).

Beyond the pollution hazard created by animal waste, intensive animal production facilities have created public outrage over their environmental impact. Outrage over environmental risk often is a complex mixture of protest against odor and pollution combined with frustration over aesthetic and demographic changes in local communities. Animal production facilities may enjoy favorable zoning because of their farm status, but intensive production facilities may have an appearance and aesthetic more typically associated with heavy industry than farming. In many parts of the United States, a dozen or more intensive production facilities may cluster in a particular community. Clustering can create a significant increase in heavy truck traffic, resulting in congestion and increased highway expenses. As noted previously, communities may be poorly prepared for the influx of a Spanish-speaking workforce. Although these are not strictly environmental problems, environmental regulation may be the only outlet for local community members to express anxiety and frustration over these negative impacts on their communities (DeLind, 1995; Purvis, 1998).

LIVESTOCK ISSUES IN CONTEMPORARY SOCIETY

The preceding overview is a very brief summary of challenges facing the livestock sector. Each of these challenges involves elements of risk. Structural transformation of animal agriculture poses financial risk to animal producers. The industry's response to financial risks has led to the intensive production systems that are the primary focus of research on livestock well-being today. Intensive, mechanized livestock production systems in the pork and poultry sector almost entirely have replaced extensive family-run operations and diversified farms. These new production systems are pejoratively referred to as "factory farms." Like factories, they depend heavily on capital investment, technology, and a mix of professional managers and low-wage laborers. The technology deployed in intensive livestock production, however, gives rise to both real and perceived risks associated with food safety and environmental quality. Food safety and environmental risks are borne by food consumers and residents of rural communities. As these risks emerge and are recognized, more people see themselves as affected by, and having, an interest in the choice of methods for animal production. As consumers, rural residents and animal welfare advocates defend interests affected by the deployment of technology in animal production. Their involvement creates uncertainty about the future economic and regulatory environment for livestock producers. This, in turn, introduces still more financial risk into their planning horizon, as both capital investments and profitability can be affected by food safety or environmental or animal welfare regulations.

Many questions remain, however, after placing the issue of animal welfare alongside those of structural transformation, food safety, and environmental impact. Why have these particular risk issues emerged in concert? What should be done about them? Who are the key actors—government, citizen activists, animal producers, other agribusiness and food system firms, or scientists—who should take responsibility in responding to this complex situation, and what are their respective roles? The answer to the first of these questions is particularly sensitive because it advances a given problematization of animal welfare within the nexus of livestock production risks. *Problematization* is a term used by Foucault (1984) to indicate the way that material realities intersect with cultural beliefs to create a particular conceptualization of a situation as problematic, as a source of anxiety or irritation, and as needing some sort of response. Any diagnosis of the problem situation will involve a mixture of science and philosophy. Although such diagnoses or problematizations have a profound influence on the type of research that is developed to address issues associated with animal welfare, their complexity, scope, and philosophical dimensions make them difficult to address using standard scientific methods.

It is possible to contrast the two themes that emerge in alternative conceptualizations of the problem situation for the welfare of agricultural animals. On the one hand, there are analysts who diagnose the emergence of multiple risk issues associated with animal production as a symptom of industrialization. The metaphor suggests that industrialization is a disease and that responses either must treat the symptoms or seek a cure. On the other hand, there are analysts who take these risk issues to be characteristic of postindustrial society. For them, a response to these risks demands the innovation of new institutions that will accommodate the diversity of interests affected by animal production. The difference between these two ways of conceptualizing the multiplicity of risks involved in livestock production easily might be overstated. Each overlaps the other in many respects, and one can find elements in the work of most analysts that fit either approach. Nevertheless, it is useful to overdraw the distinction between an industrial and a postindustrial analysis to discuss the indications and possibilities of each.

Industrial Paradigm

The model of industrial society that has held sway in Europe, North America, and in certain regions of Asia for at least 200 years assumes that human social relations are progressively organizing processes of industrial production. Industrial production organizes labor, capital, and technology into the most efficient configuration for production of commodity goods for human consumption.

The process of organization for efficient production has at least two basic sources of dynamism. One is the historical shift from a social structure characterized by landed wealth, hereditary aristocracy, and feudal obligation to the current system of capital markets, centralized states, and wage labor. The other is the growth of science and the emergence of technology that introduces a seemingly endless stream of innovations calling for constant reorganization of production in pursuit of increased efficiency. In industrial society, social problems are thought to have their roots in conflict that emerges out of individuals' need to secure a steady stream of monetary income. In a simplistic portrayal of this conflict, people are organized into social classes according to whether their income derives from capital returns and profits or compensation for labor. Political alignments evolve out of these conflicts, as the interests of capital favor laissez-faire politics and open markets, but the interests of labor favor state control of capital and industrial regulation. In their crudest form, the politics of industrial society reduce to capitalism versus socialism.

Obviously, animal production is undergoing a process of reorganization and technological transformation parallel to processes that occurred much earlier for other sectors of the economy. However, the paradigm of industrialization plays a role in framing the problems of animal agriculture in several ways. One way of conceptualizing this process sees the transformation from village life and agrarian society as deeply problematic—a kind of disease in itself. Traditional societies are thought to embody healthful and appropriate social relations that are broken down in the process of industrialization. Alienating and exploitative relationships are thought to emerge as naked economic exigencies replace the traditional moralities of family and village life. The idea implicit here is that people may have been poor in the preindustrial world but were not exposed to the risks of poisoned food or water, and their animals were not abused.

What must be done, on this view, is to restore the more proper and natural social relations of the past without sacrificing the beneficial aspects of modern technology. The vision that, to varying degrees, may underlie many communitarian and socialist beliefs is particularly influential with respect to agriculture. It becomes possible to see the decline of family farming as emblematic of all that has gone wrong in the industrial age (Thompson, 1998).

These are, of course, extremely broad characterizations that portray the application of the industrial paradigm in oversimplified terms. For example, many who sharply criticize industrialization are not inclined to such a favorable interpretation of agrarian society or family farms. Fink (1998) found family farmers every bit as insensitive to the plight of the working classes as are industrial capitalists. However, Thu and Durrenberger (1998), along with the other contributors to their book, combined a leftist orientation to the politics of industrialization with expressions of solidarity for the decline of the family farm. Maurer (1995) analyzed both health-oriented and animal welfare-oriented vegetarianism in similar terms, as did

M. W. Fox (1997). Daniel (1993) also placed the animal welfare issue squarely into a context defined primarily by the expanding power of corporate capital and the declining role of small, family farms.

Tweeten (1993) laid out an alternative way to frame the problems of industrialization in explicit fashion. Tweeten saw structural change in agriculture as a progressive process in view of its capacity to provide a higher quality of life for a vastly increased number of people. Nevertheless, the process of industrialization does not occur without cost or in a smooth and seamless way. At any given time, there will be losers as well as elements in the existing organization of labor, capital, and commodity markets that could be made more efficient. As such, there are two possible responses to complaints voiced in response to industrialization. One is to conclude that those who complain about food safety, environmental impact, or animal welfare—as well as those farmers who go under because of structural transformation—are losers. When viewed from the perspective of society as a whole, the benefits of industrial technology may more than offset these losses. The other possibility is that the costs borne by consumers, animals, and the environment are externalities. They are real costs that should have been taken into account in the drive toward more efficient production but were omitted because of an imperfection in the structure of labor, capital, or commodity markets. If so, then there may be a proper role for government regulation that will lead to an internalization of these costs and their eventual reflection in the prices paid for animal products.

Tweeten (1993), however, also argued that anyone who sees animal welfare as a market failure has rejected traditional Judeo-Christian values. He bases this claim on the assumption that the Judeo-Christian tradition in ethics limits moral consideration to other human beings. Under this assumption, harm to animals is not morally significant and should not be reflected in the balance of costs and benefits associated with an animal production system. Thus, Tweeten rejected the possibility of regulating animal production in a way that would force producers to internalize costs borne by animals in the form of suffering and deprivation. It is worth noting that this aspect of Tweeten's argument is logically separable from his general analysis of industrialization.

Because there are a number of alternative ways to understand the link between ethics and the welfare of livestock (Fraser, 1999), it should be possible to interpret adverse impact on animals as a "cost" that would be justified only under a framework that weighs all costs and benefits equitably.

Postindustrial Paradigm

The idea of "risk society" is a construct intended to conceptualize human social and political relations in a broad and comprehensive way. Beck (1992) offered it as an alternative to the conceptualizations of industrial society that are implicit in many forms of social organization and political action and that serve as the explicit foundations for social and political theory. In Beck's formulation, the industrial paradigm has been eroding steadily over the last 50 years. Although certainly it still is true that capital and labor find themselves at odds over many aspects of industrial production, Beck argued that no longer is it plausible to see this opposition as the root of all social problems. Rather than having the form of an industrial class society where sources of income define virtually all economic and political issues, Beck argued that the advanced democracies are becoming postindustrial risk societies. In a risk society, people place one another at risk in diverse and extraordinarily complex ways.

Certainly, conflicts over sources of income serve as one of the ways to be at risk or, through one's actions, to place someone else at risk. Beck (1992), however, argued that this relationship is only one of many that define social issues today. There are, in fact, many sources of potential instability in peoples' lives in contemporary society, including changes in gender and family roles. In particular, people are placed at risk from unintended consequences of technologies introduced either to increase efficiencies in the production process or to mitigate other risks. Beck's point is that these risk relations tend to be discrete and unstructured. One may be placed at risk by the actions of one's spouse, the industrial plant next door, or technological developments taking place around the globe. Indeed, one may place oneself at risk through certain consumption activities.

An approach that sees all issues in terms of left and right or that assumes consistent political solidarity based on one's economic position will founder with such issues. There is no clear way to align all the interests at risk in contemporary society in a coherent and stable way. Furthermore, any adequate response to risk demands that those who take themselves to be at risk must have confidence in the actors and the processes that are undertaken in making the response. Lacking such confidence, the feeling that one is at risk fails to be alleviated, whatever the likelihood that a hazard actually will materialize. Thus, responses to risk must involve potentially affected parties and their spokespersons. Thus far, institutions for accomplishing this involvement and engendering trust are highly experimental and ineffective. Beck (1992) recommended that those beset with risk-oriented problems should undertake negotiations and political action designed to develop new institutions that make the creation and maintenance of trust their highest priority. Involvement of affected and interested parties is presumed to be a prerequisite.

There are analysts of the animal welfare issue who have begun to apply the type of postindustrial approach suggested by Beck (1992). Kunkel (2000) offered an integrated analytic approach to structural transformation, food safety, environment, and animal welfare issues in animal agriculture that is derived, in part, from Beck's work. He noted that any successful approach to these issues will need to be sensitive to the interests of all affected parties. Kunkel argued that each of these challenges facing animal producers presumes a specific context and that solutions demand integrative research sensitive to diverse value perspectives. Verbeke and Viane (2000) discussed animal welfare and food safety as dimensions of "consumer concern." Their point was not to displace animal interests with economically driven consumer interest. Rather, "consumer concern" was a phrase offered by Rippe (2000) and Brom (2000) to indicate a forward-looking program of public debate and negotiation related to the food system that acknowledges the validity of cultural values not amenable to scientific measurement. The call to frame issues in light of consumer concern was an attempt to develop institutions that will allow a broader set of perspectives to influence the evolution of the food system (Hamilton, 1996; Stamen & Brom, 2000).

ROLE OF ANIMAL WELFARE SCIENCE:
A CONCLUDING NOTE

As indicated, the framing of problems and risks in livestock production is a complex business. The development of a problematization for livestock production blends what is known about biological processes and socioeconomic systems with elements of common sense and judgments about how people tend to see themselves, their problems, and the realistic chances of doing anything about them. If most people tend to see their problems in terms of the industrial paradigm, then solutions that radically depart from that paradigm are unlikely to find an audience. Yet, there are reasons to think that the industrial paradigm may not be the most propitious way of framing the nexus of issues facing livestock production, including animal welfare. As Stricklin and Swanson (1993) argued, animal welfare science does not support the belief that intensive methods associated with industrial animal production are unilaterally inimical to animal welfare. Industrial technology both helps and harms animal interests. As such, it is difficult to argue a positive association between welfare and family farming.

The industrial paradigm also provides a rationale for measuring the impact of current production methods on the well-being of livestock. Such measurements could be used to justify regulations or to defend a production system that met standards deemed to indicate acceptable levels of animal welfare. Clearly, animal welfare science might be deployed as a basis for making such measurements, but it is important to recognize that valid scientific measures of welfare, in themselves, would not resolve the problems associated with industrialization. For one thing, scientific measurements would need to be combined with value judgments to derive a standard of acceptable welfare. For example, a cost–risk–benefit analysis might be conducted to (a) discover whether benefits of existing production systems offset compromises to the welfare of animals or (b) determine the cost of regulations intended to improve animal welfare. The persuasiveness of the cost–risk–benefit approach depends, however, on the assumption that one is making net improvements in efficiency. Because a decline in one dimension could off-set an improvement in another, one also must include costs and risks associated with structural transformation, food safety, and environmental quality. All of these interests must go forward together. Like animal welfare, each of these dimensions involves qualitative and contentious value judgments. Thus, it is doubtful that this approach will produce uncontested or noncontroversial recommendations for live-stock producers or policymakers.

The animosity associated with attempts to measure the efficiency of livestock production is not conducive to the formation of public trust in science. To the extent that Beck (1992) was right in saying that people see their problems less and less in the terms offered by the industrial paradigm and more and more in terms of risk and trust, the use of science alone to justify industrial practices may exacerbate feelings of mistrust. The postindustrial paradigm suggests that welfare science will be more successful in addressing problems if affected parties can understand clearly how results are being used to mitigate the risks and harms that concern them. This suggests that, whenever possible, research should be conducted with participation and involvement by interested parties—with an eye toward the communication and integration of results into an open process of debate and decision in every case. Having precise and rigorously defensible measures of welfare may be less valuable than having a way to convince all interested parties that a particular strategy represents an improvement over the status quo.

As Curtis (1994) and Kunkel (2000) noted, successful welfare science for agricultural animals, in any case, will demand an integrative approach sensitive to the broad array of problems and constraints facing livestock producers. This demands sensitivity to empirical research on the economic, food safety, and ecological dimensions of livestock farming as well as a critical and reflective approach to the way that issues in livestock production are thought of as problematic. The interpretation of risk will play a crucial role in the problem formulation. An affected party's confidence in the actors and information presented to them contribute in a particularly significant way to the sense of being at risk. Assumptions commonly made in connection with the observation that animal agriculture is undergoing a process of industrialization may lead analysts to neglect opportunities to build confidence.

REFERENCES

Beck, U. (1992). *Risk society*. London: Sage.

Brom, F. W. A. (2000). Food, consumer concerns, and trust: Food ethics for a globalizing market. *The Journal of Agricultural and Environmental Ethics, 12,* 127–139.

Brouwer, E. (1998). Sheep to cows to man: A history of TSEs. In S. C. Ratzen (Ed.), *The mad cow crisis: Health and the public good* (pp. 26–34). New York: New York University Press.

Council for Agricultural Science and Technology. (1992). *Preparing U.S. agriculture for global climate change* (Rep. No. 119). Ames, IA: Author.

Cronon, W. (1993). *Nature's metropolis*. New York: Vantage.

Curtis, S. (1994). Farm animal welfare: Obligations, realities and compromises. In *Agricultural ethics: Issues for the 21st century* (ASA Special Publication No. 57; pp. 19–24). Madison, WI: American Society of Agronomy.

Daniel, P. (1993). Technology and ethics in agriculture. *Journal of Agricultural and Environmental Ethics, 6*(Suppl. 1), 52–59.

DeLind, L. B. (1995). The state, hog hotels and the "right to farm." *Agriculture and Human Values, 8*(4), 34–44.

Donham, K. J. (1998). The impact of industrial swine production on human health. In K. M. Thu & E. P. Durrenberger (Eds.), *Pigs, profits and rural communities* (pp. 73–83). Albany: State University of New York Press.

Fink, D. (1998). *Cutting into the meat packing line: Workers and change in the rural Midwest*. Chapel Hill: University of North Carolina Press.

Fite, G. C. (1981). *American farmers: The new minority*. Bloomington: Indiana University Press.

Foucault, M. (1984). Polemics, politics and problematizations. In P. Rabinow (Ed.), *The Foucault reader* (pp. 381–390). New York: Pantheon.

Fox, M. A. (1999). *Deep vegetarianism*. Philadelphia: Temple University Press.

Fox, M. W. (1997). *Eating with conscience: The bioethics of food*. Troutdale, OR: New Sage Press.

Fraser, D. (1999). Animal ethics and animal welfare science: Bridging the two cultures. *Applied Animal Behavior Science, 65,* 171–189.

Fraser, D., Mench, J., & Millman, S. (2001). Farm animals and their welfare in the year 2000. In D. J. Salem & A. N. Rowan (Eds.), *The state of the animals 2001* (pp. 87–99). Washington, DC: Humane Society Press.

Gebremedhin, T. G., & Christy, R. D. (1996). Structural changes in U.S. agriculture: Implications for small farms. *Journal of Agricultural and Applied Economics, 28,* 57–66.

Grandin, T. (Ed.). (2000). *Livestock handling and transport* (2nd ed.). Wallingford, England: CAB International.

Hamilton, R. (1996). Consumer sovereignty as ethical practice in food marketing. In B. Mepham (Ed.), *Food ethics* (pp. 138–153). New York: Routledge.

Jackson, L. L. (1998). Large-scale swine production and water quality. In K. M. Thu & E. P. Durrenberger (Eds.), *Pigs, profits and rural communities* (pp. 103–119). Albany: State University of New York Press.

Kunkel, H. O. (2000). *Human issues in animal agriculture*. College Station: Texas A&M University Press.

Letson, D., Gollehon, N., Breneman, V., Kascak, C., & Mose, C. (1998). Confined animal production and groundwater protection. *Review of Agricultural Economics, 20,* 348–364.

Levine, J. (1986). Hearts and minds: The politics of diet and heart disease. In H. M. Sapolsky (Ed.), *Consuming fears: The politics of product risks* (pp. 40–79). New York: Basic Books.

Martin, L. L. (1997). Production contracts, risk shifting and relative performance payments in the pork industry. *Journal of Agricultural and Applied Economics, 29,* 267–278.

Maurer, D. (1995). Meat as a social problem: Rhetorical strategies in contemporary vegetarian literature. In D. Maurer & J. Sobel (Eds.), *Eating agendas: Food and nutrition as social problems* (pp. 143–163). New York: de Gruyter.

Purvis, A. (1998). An industrializing animal agriculture: Challenges and opportunities associated with clustering. In S. A. Wolf (Ed.), *Privatization of information and agricultural industrialization* (pp. 117–150). Boca Raton, FL: CRC Press.

Rippe, K. P. (2000). Novel foods and consumer rights: Concerning food policy in a liberal state. *Journal of Agricultural and Environmental Ethics, 12,* 71–80.

Schell, O. (1984). *Modern meat.* New York: Random House.

Sere, C., & Steinfeld, H. (1996). *World livestock production systems* (Animal Production and Health Paper 127). Rome: Food and Agriculture Organization of the United Nations.

Sinclair, U. (1981). *The jungle.* New York: Bantam. (Original work published 1906)

Stamen, J., & Brom, F. W. A. (2000). Proposal for a transatlantic platform for consumer concerns and international trade. *Journal of Agricultural and Environmental Ethics, 12,* 207–214.

Stricklin. W. R., & Swanson, J. C. (1993). Technology and animal agriculture. *Journal of Agricultural and Environmental Ethics, 6*(Suppl. 1), 67–80.

Thompson, P. B. (1997). Science policy and moral purity: The case of animal biotechnology. *Agriculture and Human Values, 14,* 11–27.

Thompson, P. B. (1998). *Agricultural ethics: Research, teaching and public policy.* Ames: Iowa State University Press.

Thu, K., & Durrenberger, E. P. (1998). Introduction. In K. M. Thu & E. P. Durrenberger (Eds.), *Pigs, profits and rural communities* (1–20). Albany: State University of New York Press.

Tweeten, L. (1993). Public policy decisions for farm animal welfare. *Journal of Agricultural and Environmental Ethics, 6*(Suppl. 1), 87–104.

Verbeke, W. A., & Viane, J. (2000). Ethical challenges for livestock production: Meeting consumer concerns about meat safety and animal welfare. *Journal of Agricultural and Environmental Ethics, 12,* 141–151.

Wallerstein, I. (1974). *The modern world system I: Capitalist agriculture and the origins of the European world economy in the 16th century.* New York: Academic.

JOURNAL OF APPLIED ANIMAL WELFARE SCIENCE, 4(3), 207–221

Animal Welfare Issues in the Poultry Industry: Is There a Lesson to Be Learned?

Ian J. H. Duncan

Department of Animal and Poultry Science
University of Guelph, Canada

Many of the conditions in which poultry live and the procedures to which they are subjected compromise their welfare. This article describes these welfare problems in the hope that they may serve as warnings to the rest of animal agriculture, which then might take steps to avoid the same pitfalls. The article discusses poultry welfare problems under the headings of battery cages for laying hens, forced molting, disposal of spent laying hens, fast growth problems in meat poultry, catching and transportation, food restriction of broiler breeders, hyperaggressiveness in broiler breeder males, elective surgeries, and water-bath stunning.

Poultry are kept more intensively, both for meat and for egg production, than are animals in any other sector of animal agriculture. *Intensively* refers to the numbers of animals per production unit, the degree of crowding to which they are subjected, and the artificiality of the environment in which they are kept. As this article reveals, the poultry industry also has many animal welfare problems. The aim of this article is to describe these poultry welfare problems and how they have arisen and thus erect some warning flags for the rest of animal agriculture.

WELFARE PROBLEMS OF LAYING HENS

Battery Cages

The battery cage system for laying hens was one of the first intensive husbandry systems to come under criticism on animal welfare grounds (Harrison, 1964; Command Paper 2836, 1965). The criticisms have continued unabated (Singer,

---

Requests for reprints should be sent to Ian J. H. Duncan, Department of Animal and Poultry Science, University of Guelph, Guelph, Ontario, Canada N1G 2W1. E-mail: iduncan@uoguelph.ca

1990; A. J. F. Webster, 1995). In Europe, the movement against traditional cages has been so great that the European Union has approved a directive (Commission of the European Communities, 1999) to ban cages. This directive prohibits traditional battery cages from January 1, 2012. Only furnished or enriched cages will be allowed from 2012. In addition, no battery cages may be installed from January 1, 2003, and standards are set for all "alternative systems."

Despite all the criticisms of traditional cages, it should be remembered that there are some advantages to keeping hens in cages and that certain of these actually are welfare benefits. For example, the increased hygiene achieved through separating the hen from her feces results in a much lower incidence of diseases where the infectious agent is spread via the droppings (Duncan, 2000). In addition, the small group size in cages is much closer to the group size that hens prefer (Dawkins, 1982; Hughes, 1977; Lindberg & Nicol, 1993, 1996).

Nevertheless, there are many welfare problems associated with cages. Possibly, the biggest problem is the lack of a nesting site. Once nesting behavior triggers internal hormonal factors (Wood-Gush & Gilbert, 1975) hens are strongly motivated to find a suitable nesting site (Duncan & Kite, 1989) and will work hard to obtain one (Follensbee, Duncan, & Widowski, 1992). In addition, many hens, particularly those of light hybrid strains who are commonly used in North America, show symptoms of severe frustration in the prelaying period in cages (Mills, Duncan, Slee, & Clark, 1985; Mills, Wood-Gush, & Hughes, 1985; Wood-Gush, 1972).

The lack of space in battery cages reduces welfare by preventing hens from adopting certain postures—such as an erect posture with the head raised—and performing particular behavior, such as wing flapping (Dawkins & Hardie, 1989; Nicol, 1987a, 1987b). Apart from nesting behavior, the behavior systems that have been investigated the most are perching and dust bathing, and there is evidence that their prevention reduces welfare (Duncan, 2000). The lack of movement and exercise also contributes to bone weakness in spent laying hens (Leeson, Diaz, & Summers, 1995). In addition, of course, the lack of space means that hens are crowded together. All of the indications are that welfare is decreased (Hughes, 1975; Mashaly, Webb, Youtz, Roush, & Graves, 1984) at cage densities used in North America (450 cm^2 per bird in Canada and 350 cm^2 per bird in the United States).

The decision already has been made in Europe that the welfare problems of cages more than outweigh the advantages. What will happen in North America? It seems unlikely that there will be a ban on cages in the near future. However, cages may have to be modified to make them more welfare-friendly. There already is some interest in Canada in trying out some of the new European furnished cages (Appleby, 2000). There certainly will be an increasing market for noncage eggs.

The lesson here for other sectors of animal agriculture would seem to be that a look toward Europe is sometimes useful. In the field of animal welfare, North America seems to lag behind Europe by about 8 to 10 years. In the United States, for example, rodents and birds only now are being considered for inclusion in the

Animal Welfare Act, and birds are not included under the Humane Slaughter Act. In contrast, Europe is phasing out battery cages. Dry sow stalls already have been banned in some European countries, and a European Union phaseout is in progress. Wise North American producers should be taking note of these developments and planning their future husbandry systems accordingly.

Forced Molting

After laying eggs for about 1 year, commercial laying hens start to become photorefractory (unresponsive to long days), egg production starts to fall, and eggshell quality decreases. The bird's skeleton has been depleted of calcium through producing many eggshells and is fragile; the hen often is overweight. If hens were then exposed to short days, say of 8 hr, they would gradually go out of reproductive condition and would molt naturally. However, natural molting is a slow process, and there would be a wide range of times within a flock of hens for individual hens to complete the molt. Currently, the poultry industry is not prepared to accept this extended loss of production. At about 74 weeks of age, therefore, hens are sent to slaughter as "spent laying hens" (in Canada) or are "force-molted" (in the United States) to speed up the molting process and get the hens back into reproductive condition for a 2nd and sometimes a 3rd laying year.

Forced molting programs usually involve withholding feed for 10 to 14 days and simultaneously reducing day length (Leeson & Summers, 1991; North & Bell, 1990). Forced molting shortens the period of nonproduction to about 8 weeks but results in a huge increase in stress and suffering. A rather crude indicator of reduced welfare is increased mortality. During forced molting, mortality increases dramatically. Duncan and Mench (2000) cited Bell, who summarized molting results from 353 U.S. flocks during 1997 and 1998 and found that mortality typically doubled during the 1st week of molt, then doubled again during the 2nd week.

Apart from mortality, however, the evidence suggests that hens suffer enormously during forced molting. Hunger is an extremely powerful motivation, and chickens have evolved to forage and consume food throughout the day (Savory, Wood-Gush, & Duncan, 1978). Consequently, deprivation of food acts as a drastic stressor. Food deprivation results in a classical physiological stress response (Mench, 1991). Frustration of feeding leads to signs of extreme distress such as increased aggression (Duncan & Wood-Gush, 1971) and the formation of stereotyped pacing (Duncan & Wood-Gush, 1972). Extremely hungry birds also show stereotypic pecking at objects such as feeders (Hocking, Maxwell, & Mitchell, 1996; Kostal, Savory, & Hughes, 1992; Savory & Maros, 1993). In an experiment in which hens were deprived of food for 3 days, A. B. Webster (1995) found that cage pecking increased by a factor of 3 and feather pecking by a factor of 8. In a later study designed to simulate forced molting, A. B. Webster (2000) deprived hens of feed for 21 days. Hens subjected to this deprivation at first showed increased aggression and nonnutritive pecking suggestive of severe frustration and extreme hunger and, later, inactivity suggestive of debilitation (Duncan & Wood-Gush, 1971, 1972).

The rest of animal agriculture would do well to be wary of extreme procedures. For example, segregated early weaning of piglets at 10 to 14 days of age is an extreme procedure that very likely reduces welfare.

Disposal of Spent Laying Hens

Of all the animal welfare problems faced by the poultry industry today, the disposal of spent laying hens probably is the most serious (Newberry, Webster, Lewis, & Van Arnam, 1999). A spent laying hen is a hen at the end of her productive life. It is usually arranged—by manipulating body weight and day length—for laying hens to start laying eggs at approximately 20 weeks of age. They lay eggs for about 1 year, at which point decreasing egg numbers and egg-shell quality mean that it is no longer profitable to continue. When the hens are about 74 weeks old, they either are sent for slaughter as spent laying hens or force-molted and kept for a 2nd laying year. The majority of hens are disposed of after this 2nd laying year; a small number of flocks may be force-molted again and kept for a 3rd laying year. No matter how many years they have been in lay, all laying hens are eventually slaughtered as spent laying hens.

It is difficult to gauge just how serious this welfare problem is. In a British survey, however, 29% of hens from battery cages were found to have freshly broken bones just before they were stunned, and most of this damage occurred as the birds were being removed from the cages (Gregory & Wilkins, 1989). Spent laying hens in North America are handled in the same way, so there is no reason to think that American statistics would be any better.

There appear to be three main reasons for this appalling statistic. First, hens kept in battery cages, even for 1 laying year, have very fragile skeletons (Knowles & Broom, 1990; McLean, Baxter, & Michie, 1986; Norgaard-Nielsen, 1990). There is such a high demand for eggshell calcium in modern laying hens that cortical as well as medullary bone is used as a source of calcium, which results, at the end of 1 laying year, in easily broken bones (Leeson et al., 1995). The bone weakness is exacerbated by lack of exercise in cages (Leeson et al., 1995). Second, traditional battery cages are poorly designed for the removal of hens. Small doors to the cages result in hens getting limbs caught as they are being removed. Modern European cages are designed so that the whole front opens up, resulting in a much lower risk of damage (Tauson, 1980, 1989). Third, spent laying hens are worth very little, and so no effort is made to handle them carefully. The combination of these three factors—fragile skeleton, poorly designed cage, and low value—results in an unacceptably high injury level.

Because only a few processing plants are prepared to accept spent hens, the problem is made worse by the often long journeys to slaughter. This means that injured hens may be in pain for long periods. When they reach the processing plant, their problems continue. The tetany and muscular spasms that accompany electrical stunning lead to further bone breakage in spent hens with their fragile skeletons (Gregory & Wilkins, 1989). To reduce this in countries where electrical stunning is practiced, there is a tendency on the killing line to reduce the intensity of the electrical stun. This increases the risk that some hens will not be stunned properly before slaughter. If they are not stunned properly, they do not assume the characteristic posture during tetany and face a bigger risk of missing the automatic cutting machine. Thus, they run a higher risk of entering the scalding tank alive and conscious (Duncan, 1997).

The disposal of spent laying hens is proving to be an intractable problem. It has been suggested that perhaps the hens should be killed in the barn and composted, but this gives a very wasteful image. The most humane way would be to kill the birds while still in the cages, say by gassing. However, the problem then is a mechanical one of removing bodies stiffened by rigor mortis from the cages. Because there also might be safety issues for humans associated with using a poisonous gas, there is some interest in developing a portable carbon dioxide gas-stunning and killing cabinet (A. B. Webster, Fletcher, & Savage, 1996) into which hens could be placed on removal from the cages.

The problem in North America could be eased by switching from light body-weight hybrid strains of laying hen to medium body-weight hybrid strains that are more robust and have more value as spent hens. However, these medium hybrid strains have been bred to produce brown eggs, and there is little current demand for brown eggs in North America.

The lesson here for other sectors of the livestock industry is that the welfare of "low-value" animals is at great risk and that safeguards need to be established for their protection. Probably, the animals most at risk are cull dairy cows, cull boars, cull sows, and any animals with reduced value due to some market quirk. The cost of humanely disposing of low-value animals must be factored into the costs of production for that commodity. For example, the only way found to protect the environment from the dumping of worthless used automobile tires is to add a charge for their disposal when they are bought as new tires. Similarly, the cost of humanely disposing of spent laying hens should be added to the price of eggs. Even if the cost were as high as $1 per hen, this only would amount to an additional cost of about .33 cents per egg.

WELFARE PROBLEMS OF MEAT POULTRY

Fast Growth Problems in Meat Poultry

There are reports of an increasing incidence of conditions such as skeletal deformities and ascites that accompany fast growth in meat strains of poultry (Julian, 1998; Leeson et al., 1995). *Ascites* is the condition that occurs when a rapidly growing bird has insufficient heart–lung capacity to supply all of the soft tissues with oxygenated blood. This leads to an increase in blood pressure, dilation and hypertrophy of the right ventricle, and leakage of serous fluid into the body cavity (Julian, 1998). Because these conditions cause the birds to suffer, they are both welfare and production problems. For example, there is evidence that these skeletal deformities are painful. When given a choice between two feeds—one of which contained an analgesic—broilers with gait abnormalities consumed more of the drugged feed than did broilers with no lameness. Moreover, the walking ability of lame birds was improved by this self-administered treatment (Danbury, Weeks, Chambers, Waterman-Pearson, & Kestin, 2000). In another experiment, the amount of spontaneous movement shown by male turkeys was increased greatly by the administration of a drug that reduces pain and inflammation in arthritic joints. These turkeys were later shown to have degenerative lesions of the hip joints (Duncan, Beatty, Hocking, & Duff, 1991).

The increasing incidence of fast growth problems such as these in meat strains of poultry indicate that we are reaching the biological limit of growth and that it is a mistake to think we can go on and on selecting for increased growth rate without costs to the bird. It also is a mistake to think that we somehow can find an environmental or nutritional solution to these problems. The long-term solution will be a genetic one. McMillan (2000) developed a computer simulation in which the effects of four different genetic selection procedures for growth rate and against incidence of ascites were compared. All of these procedures resulted in increased growth rate, but they also resulted in an increase in the level of ascites.

The primary breeding companies should heed this warning, stop selecting for increased growth, and try to add value to their strains of bird by some other means. Because the breeding companies assume that this would put them at a competitive disadvantage with their rivals, they, of course, are reluctant to stop selecting for fast growth.

The other sectors of the livestock industry should be wary of intense selection for fast growth. The swine industry, which is following the same pattern of genetic selection as the meat chicken industry, already is running into these problems (Grandin & Deesing, 1998). There also may be problems of selecting intensively for leanness in both swine and cattle (Grandin & Deesing, 1998).

Food Restriction of Broiler Breeders

Broiler breeders—the parent stock who produce broilers—have the same huge appetites as their progeny and have to be maintained on very severe food restriction so that they are able to reproduce. If allowed free access to food, they soon become obese and suffer from all of the problems of obesity, including low fertility and reduced life expectancy (Leeson & Summers, 2000; Renema & Robinson, 2000). Food restriction is carried out for a very good reason: to keep the birds in good reproductive condition and prevent them becoming obese, a condition that itself reduces welfare (Renema & Robinson, 2000). However, these food-restricted birds exhibit behavioral symptoms that indicate greatly reduced welfare (Mench & Falcone, 2000; Savory, 1989). Once again, the producer is in a dilemma. If the birds are fed to appetite, they will become obese and long-term welfare will be reduced; if they are restricted, then they show symptoms of hunger and extreme distress. It may be possible to alleviate the problem in the short term by diluting the diet with nonnutritive substances such as cellulose (Savory, Hocking, Mann, & Maxwell, 1996; Zuidhof et al., 1995). However, in the long term, the solution must be to develop parent stock with smaller appetites. When the primary breeding companies stop selecting for growth rate, perhaps this problem will be resolved.

The swine industry already has encountered this problem, with breeding sows having to be kept severely food restricted during gestation. There is evidence that it is the food restriction during gestation as much as the stall environment that leads to the development of stereotyped oral activities in sows in dry stalls (Appleby & Lawrence, 1987). In this case, it seems to be too late to learn from the poultry industry mistake.

Hyperaggressive Behavior in Broiler Breeder Males

A new problem emerged in the poultry industry in the 1990s. An increasing number of reports described broiler breeder males being very aggressive toward females (Mench, 1993). This is highly unusual because male domestic fowl dominate females passively and seldom show any overt aggression toward them (Wood-Gush, 1956). Because females are being harassed, badly injured, and even killed by males, this is a welfare as well as a production problem. Investigation has shown that most broiler breeder males of various strains are very aggressive toward females (Millman, Duncan, & Widowski, 2000). This cannot be explained in terms of a general increase in aggression, because game fowl males who have been bred for fighting and who are much more aggressive toward other males than are broiler breeder males (Millman & Duncan, 2000c) show little, if any, aggression toward females (Millman & Duncan, 2000b). The aggression is not caused by the males being food restricted during either the rearing phase (Millman & Duncan, 2000b) or the adult phase (Millman et al., 2000) and almost certainly has a genetic basis. It also has been shown that broiler breeder males are deficient in certain elements of courtship behavior (Millman et al., 2000). The result is that the females do not react appropriately when the males approach but move away and avoid them (Millman & Duncan, 2000a).

Now, we only can speculate how this problem has arisen. It is not clear whether the courtship deficiency and the hyperaggressiveness are separate or linked problems. It may be that these traits are linked genetically to some production trait, such as broad-breastedness, for which the breeding companies have been selecting. On the other hand, poor fertility is a problem with broiler breeders, particularly toward the end of the breeding year. It is widely thought within the poultry industry that this is due to decreased libido. However, it is actually due to the males being unable to achieve cloacal contact with the females because of their conformation (Duncan, Hocking, & Seawright, 1990). Therefore, the breeding companies may have been selecting males who approach females very quickly in the mistaken belief that they are very sexy. In fact, these males are aggressive.

The rest of animal agriculture should be very wary of intensive genetic selection for a particular trait without taking account of the animal's total biology.

GENERAL WELFARE PROBLEMS
IN THE POULTRY INDUSTRY

Elective Surgeries or Mutilations

The poultry industry carries out several elective surgeries routinely. For example, most laying hens in North America are beak-trimmed or debeaked. Male chicks destined for breeding are often *dubbed*; that is, they have their combs trimmed so that the comb grows as a more compact mass and has a rounded top surface less likely to be damaged in adulthood (Leeson & Summers, 2000). Male turkeys usually are *desnooded* as chicks; that is, they have the fleshy protuberance that hangs over their beaks trimmed so that the shortened snoods are less likely to be damaged in adulthood. As chicks, broiler breeder males often have the tip of the third phalanx removed from the inside toe or the two inside toes with a hot blade so that they cause less damage to the hen when mounting her in adulthood (Leeson & Summers, 2000). Many turkeys, male and female, also have this surgery as chicks so that in adulthood they cause less scratching damage if they panic and clamber over one another. Therefore, the reason for all these surgeries is to prevent damage later in life. The appendages causing the damage (the beak or the toes) are modified so that they cause fewer injuries, and the appendages likely to be damaged (the combs and the snoods) are modified so that they are less likely to be injured.

It could be argued that all these surgeries are carried out for welfare reasons—to prevent pain and injury later in life. When these surgeries have been carefully investigated, however, welfare costs have been found that should be balanced against the benefits. This is illustrated best using the example of beak trimming or debeaking.

In fact, neither of these terms—beak trimming or debeaking—is strictly accurate. When the birds are in the growing phase, one third of the upper beak can be amputated using a sharp heated blade. Alternatively, a precision machine with a laser beam or a powerful electric spark can punch a hole in the beak when the birds are still chicks, and the end of the beak sloughs off a few days later. The beak of the fowl is well innervated and contains both mechanoreceptors and nociceptors (Breward, 1984). It has been shown that when the beak is partially amputated using a hot-blade debeaker during the growing phase (6–16 weeks), the severed nerves grow back into the damaged stump and form neuromas (benign fibrous tumors), which then send spontaneous pain signals back to the brain (Breward & Gentle, 1985). This seems similar to the phenomenon that causes phantom limb pain in human amputees.

In addition, behavioral changes suggestive of acute pain have been found to occur in the 2 days following surgery. These are followed by changes indicating chronic pain that last at least 5 or 6 weeks after the surgery (Duncan, Slee, Seawright, & Breward, 1989; Gentle, Waddington, Hunter, & Jones, 1991). This neural and behavioral evidence suggests that the idea of beak trimming being a short-lived discomfort may be far from accurate; beak trimming causes a reduction in welfare through causing pain. The problem is that beak trimming is carried out for the very good reason of preventing or controlling feather pecking and cannibalism, which can cause great suffering. The evidence suggests that it is not possible to control feather pecking completely by keeping hens in other, more extensive, environments (Appleby, Hughes, & Elson, 1992), that it has hereditary characteristics (Cuthbertson, 1980; Kjaer & Sørensen, 1997; Richter, 1954), and that unintentional genetic selection may have increased its incidence (Cuthbertson, 1980).

The long-term solution to this problem undoubtedly will be a genetic one. Muir and Craig (1998) showed that it is possible to select against feather pecking and cannibalism using a kin selection method. They kept groups of closely related hens with intact beaks in cage conditions likely to stimulate feather pecking and cannibalism. Any groups showing feather-pecking damage or damage from cannibalism were eliminated from the breeding program. Through this selection procedure, Muir and Craig produced a line of birds that they claim do not require beak trimming. The challenge will be to persuade the primary breeding companies to adopt such a procedure. If a breeding company were to start selecting against feather pecking and cannibalism, it would have to relax selection on at least some of the economic traits and, thus, put itself at a competitive disadvantage. Naturally, the breeding companies are reluctant to do this.

There has been little investigation into the welfare costs of the other elective surgeries carried out on poultry. However, Gentle and Hunter (1988) produced neuronal evidence suggesting that detoeing may be painful at the time of amputation but is less likely than beak trimming to be followed by chronic pain. There also are reports that toe clipping turkeys depresses growth rate and increases mortality (Newberry, 1992; Owings, Balloun, Marion, & Thomson, 1972), which is highly suggestive of decreased welfare.

Chopping off parts of animals that create a problem in modern intensive husbandry systems seems such a crude solution. The surgeries all are performed without anesthesia or analgesia and, at the very least, will cause some acute pain.

Other sectors of animal agriculture would do well to consider elective surgeries very carefully and be wary of introducing new surgeries such as tail docking of dairy cows. Are they all necessary? Are there alternative solutions? Could the surgeries be more humane?

Catching and Transportation

Of all the things we do to our animals on the farm, the things we do to them in the 24 hr before they are slaughtered reduce their welfare the most (Duncan, 1994). The surveys that have been carried out during catching and transportation have shown that this is just as true for poultry species as for other farm livestock (Broom & Knowles, 1989; Duncan, 1989). Birds often are injured during catching and crating, frightened by novel stimuli, stressed by disruptions to their social and physical environment throughout the catching and transportation process, and subjected to climatic extremes during transportation. Weeks and Nicol (2000) discussed these problems in detail.

A great deal of effort has gone into improving the whole catching and transportation process. For example, chicken-catching machines have been developed that pick up birds from the barn floor and very gently place them in transportation crates. This process causes much less stress and damage to the birds than does traditional manual catching (Duncan, Slee, Kettlewell, Berry, & Carlisle, 1986). In addition, transportation vehicles are being developed that monitor and control the environment of the birds to minimize stress (Mitchell, Carlisle, Hunter, & Kettlewell, 2000; Mitchell & Kettlewell, 1993). Ideally, the whole catching, transportation, and preslaughter system should be integrated with an automated catching machine placing birds in crates, modules of crates being placed on environmentally controlled trucks, and the crates moving straight into a gas-stunning unit at the processing plant (Kettlewell, Hampson, Berry, Green, & Mitchell, 2000). The challenge now is to get the poultry industry to adopt these methods; again, there will be a cost involved.

Other sectors of the livestock industry should look carefully at catching, transportation, and preslaughter management; there probably are better methods available.

Water-Bath Stunning

In most of the civilized countries of the world, poultry are stunned in a water bath before being killed by exsanguination. In Canada, more than 90% of all birds slaughtered, including meat chickens, spent laying hens, and turkeys, are stunned in this way. The exceptions are birds killed according to religious slaughter laws. In the United States, on the other hand, poultry are not included under humane slaughter laws, and a lower proportion of birds are electrically stunned.

Although water-bath stunning, when introduced about 40 years ago, represented a huge increase in welfare, it is far from ideal—not very efficient and not very humane (Duncan, 1997). When birds arrive at the processing plant, they are taken out of the crates in which they have been transported and hung by the legs on a shackle line. The line moves into the plant and over a water bath so that the birds' heads go into the water. An electrical potential between the line and the water should render every bird unconscious. However, the system contains many variables. Differences in the size of the birds, differences in the conductivity of the birds, changes in the conductivity of the water as it becomes dirty, and other variables all affect how much current travels through the birds' brains and, therefore, how well they are stunned (Duncan, 1997). There is much research going on to try to make this process more efficient and foolproof (Fletcher, 2000; Raj, 1998, 2000).

An alternative method of stunning poultry is gas or modified atmosphere stunning. This method uses the inert gas argon or a mixture of argon and carbon dioxide (Raj, 1993). It has many welfare advantages. Birds are stunned in the crates in which they have been transported, thus avoiding the stress of being shackled while conscious. Losing consciousness through anoxia is extremely quick and painless (Woolley & Gentle, 1988). There is no recovery. The birds actually are killed by anoxia before being shackled and bled. Switching from water-bath stunning to gas stunning would add a small cost to the final product. However, there are other commercial advantages. For example, the conditions for the people hanging birds on the shackles is much better as there is less noise, less dust, and more light. In addition, the workers can stand in a more ergonomically correct position. Compared with electrical stunning as it is used in Europe and Canada, gas stunning also gives a better quality product with less damage and bruising and allows for quicker further processing. Because the gas is inert, its use also means a very safe working environment (Duncan 1997; Raj, 1993).

The lesson here is that we should examine all slaughter procedures carefully; there may be better methods available.

REFERENCES

Appleby, M. C. (2000). Modifying the cage to accommodate behaviour. *Proceedings of the 21st World's Poultry Congress* (Paper S1.1.05). Montreal, Canada: World's Poultry Science Association.

Appleby, M. C., Hughes, B. O., & Elson, H. A. (1992). *Poultry production systems: Behaviour, management and welfare.* Wallingford, England: CAB International.

Appleby, M. C., & Lawrence, A. B. (1987). Food restriction as a cause of stereotypic behaviour in tethered gilts. *Animal Production, 45,* 103–110.

Breward, J. (1984). Cutaneous nociceptors in the chicken beak. *Journal of Physiology, 346,* 56.

Breward, J., & Gentle, M. J. (1985). Neuroma formation and abnormal afferent nerve discharges after partial beak amputation (beak trimming) in poultry. *Experientia, 41,* 1132–1134.

Broom, D. M., & Knowles, T. G. (1989). The assessment of welfare during the handling and transport of spent hens. In J. M. Faure & A. D. Mills (Eds.), *Proceedings of the Third European Symposium on Poultry Welfare* (pp. 79–91). Tours, France: French Branch of the World Poultry Science Association.

Commission of the European Communities. (1999). Council directive for laying down minimum standards for the protection of laying hens kept in various systems of rearing. *CEC Directive,* 1999/74/EG.

Report of the technical committee to enquire into the welfare of animals kept under intensive livestock husbandry systems (Command Paper 2836). (1965). London: Her Majesty's Stationery Office.

Cuthbertson, G. J. (1980). Genetic variation in feather-pecking behaviour. *British Poultry Science, 21,* 447–450.

Danbury, T. C., Weeks, C. A., Chambers, J. P., Waterman-Pearson, A. E., & Kestin, S. C. (2000). Self-selection of the analgesic drug Carprofen by lame broiler chickens. *Veterinary Record, 146,* 307–311.

Dawkins, M. S. (1982). Elusive concept of preferred group size in domestic hens. *Applied Animal Ethology, 8,* 365–375.

Dawkins, M. S., & Hardie, S. (1989). Space needs of laying hens. *British Poultry Science, 30,* 413–416.

Duncan, I. J. H. (1989). The assessment of welfare during the handling and transport of broilers. In J. M. Faure & A. D. Mills (Eds.), *Proceedings of the Third European Symposium on Poultry Welfare* (pp. 93–107). Tours, France: French Branch of the World Poultry Science Association.

Duncan, I. J. H. (1994). Practices of concern. *Journal of the American Veterinary Medical Association, 204,* 379–384.

Duncan, I. J. H. (1997). *Killing methods for poultry: A report on the use of gas in the U.K. to render birds unconscious prior to slaughter.* Guelph, Canada: Colonel K. L. Campbell Centre for the Study of Animal Welfare.

Duncan, I. J. H. (2000). The pros and cons of cages. *Proceedings of the 21st World's Poultry Congress* (Paper S1.1.01). Montreal, Canada: World's Poultry Science Association.

Duncan, I. J. H., Beatty, E. R., Hocking, P. M., & Duff, S. R. I. (1991). An assessment of pain associated with degenerative hip disorders in adult male turkeys. *Research in Veterinary Science, 50,* 200–203.

Duncan, I. J. H., Hocking, P. M., & Seawright, E. (1990). Sexual behaviour and fertility in broiler breeder domestic fowl. *Applied Animal Behaviour Science, 26,* 201–213.

Duncan, I. J. H., & Kite, V. G. (1989). Nest site selection and nest building behaviour in domestic fowl. *Animal Behaviour, 37,* 215–231.

Duncan, I. J. H., & Mench, J. A. (2000). Does hunger hurt? *Poultry Science, 79,* 934.

Duncan, I. J. H., Slee, G. S., Kettlewell, P. J., Berry, P. S., & Carlisle, A. J. (1986). Comparison of the stressfulness of harvesting broilers by machine and by hand. *British Poultry Science, 27,* 87–92.

Duncan, I. J. H., Slee, G. S., Seawright, E., & Breward, J. (1989). Behavioural consequences of partial beak amputation (beak trimming) in poultry. *British Poultry Science, 30,* 479–488.

Duncan, I. J. H., & Wood-Gush, D. G. M. (1971). Frustration and aggression in the domestic fowl. *Animal Behaviour, 19,* 500–504.

Duncan, I. J. H., & Wood-Gush, D. G. M. (1972). Thwarting of feeding behaviour in the domestic fowl. *Animal Behaviour, 20,* 444–451.

Fletcher, D. L. (2000). Stunning of poultry. *Proceedings of the 21st World's Poultry Congress* (Paper S3.13.02). Montreal, Canada: World's Poultry Science Association.

Follensbee, M. E., Duncan, I. J. H., & Widowski, T. M. (1992). Quantifying nesting motivation of domestic hens. *Journal of Animal Science, 70*(Suppl. 1), 50.

Gentle, M. J., & Hunter, L. N. (1988). Neural consequences of partial toe amputation in chickens. *Research in Veterinary Science, 45,* 374–376.

Gentle, M. J., Waddington, D., Hunter, L. N., & Jones, R. B. (1991). Behavioural evidence for persistent pain following partial beak amputation in chickens. *Applied Animal Behaviour Science, 27,* 149–157.

Grandin, T., & Deesing, M. J. (1998). Genetics and animal welfare. In T. Grandin (Ed.), *Genetics and the behavior of domestic animals* (pp. 319–346). San Diego, CA: Academic.

Gregory, N. G., & Wilkins, L. J. (1989). Broken bones in domestic fowl: Handling and processing damage in end-of-lay battery hens. *British Poultry Science, 30,* 555–562.

Harrison, R. (1964). *Animal machines.* London: Vincent Stuart.

Hocking, P. M., Maxwell, M. H., & Mitchell, M. A. (1996). Relationship between the degree of food restriction and welfare indices in broiler breeder females. *British Poultry Science, 37,* 263–278.

Hughes, B. O. (1975). The concept of an optimal stocking density and its selection for egg production. In B. M. Freeman & K. N. Boorman (Eds.), *Economic factors affecting egg production* (pp. 271–298). Edinburgh, Scotland: British Poultry Science.

Hughes, B. O. (1977). Selection of group size by individual laying hens. *British Poultry Science, 18,* 9–18.

Julian, R. J. (1998). Rapid growth problems: Ascites and skeletal deformities in broilers. *Poultry Science, 77,* 1773–1780.

Kettlewell, P. J., Hampson, C. J., Berry, P. S., Green, N. R., & Mitchell, M. A. (2000). New developments in bird harvesting, live haul and unloading in the United Kingdom. *Proceedings of the 21st World's Poultry Congress* (Paper S3.13.01). Montreal, Canada: World's Poultry Science Association.

Kjaer, J., & Sørensen, P. (1997). Feather pecking in White Leghorn chickens—A genetic study. *British Poultry Science, 38,* 333–341.

Knowles, T. G., & Broom, D. M. (1990). Limb bone strength and movement in laying hens from different housing systems. *Veterinary Record, 126,* 354–356.

Kostal, L., Savory, C. J., & Hughes, B. O. (1992). Diurnal and individual variation in behaviour of restricted-fed broiler breeders. *Applied Animal Behaviour Science, 32,* 361–374.

Leeson, S., Diaz, G., & Summers, J. D. (1995). *Poultry metabolic disorders and mycotoxins.* Guelph, Canada: University Books.

Leeson, S., & Summers, J. D. (1991). *Commercial poultry nutrition.* Guelph, Canada: University Books.

Leeson, S., & Summers, J. D. (2000). *Broiler breeder production.* Guelph, Canada: University Books.

Lindberg, A. C., & Nicol, C. J. (1993). Group size preferences in laying hens. In C. J. Savory & B. O. Hughes (Eds.), *Proceedings of the Fourth European Symposium on Poultry Welfare* (pp. 249–250). Potters Bar, England: Universities Federation for Animal Welfare.

Lindberg, A. C., & Nicol, C. J. (1996). Space and density effects on group size preferences in laying hens. *British Poultry Science, 37,* 709–721.

Mashaly, M. M., Webb, M. L., Youtz, S. L., Roush, W. B., & Graves, H. B. (1984). Changes in serum corticosterone concentration of laying hens as a response to increased population density. *Poultry Science, 63,* 2271–2274.

McLean, K. A., Baxter, M. R., & Michie, W. (1986). A comparison of the welfare of laying hens in battery cages and in a perchery. *Research and Development in Agriculture, 3,* 93–98.

McMillan, I. (2000). Selection for improved growth and reduced ascites syndrome incidence. *Proceedings of the 21st World's Poultry Congress* (Paper S2.4.05). Montreal, Canada: World's Poultry Science Association.

Mench, J. A. (1991). Feed restriction in broiler breeders causes a persistent elevation in corticosterone secretion that is modulated by dietary tryptophan. *Poultry Science, 70,* 2547–2550.

Mench, J. A. (1993). Problems associated with broiler breeder management. In C. J. Savory & B. O. Hughes (Eds.), *Proceedings of the Fourth European Symposium on Poultry Welfare* (pp. 195–207). Potters Bar, England: Universities Federation for Animal Welfare.

Mench, J. A., & Falcone, C. (2000). Welfare concerns in feed-restricted meat-type poultry parent stocks. *Proceedings of the 21st World's Poultry Congress* (Paper S3.3.03). Montreal, Canada: World's Poultry Science Association.

Millman, S. T., & Duncan, I. J. H. (2000a). Do female broiler breeder fowl display a preference for broiler breeder or laying strain males in a Y-maze test? *Applied Animal Behaviour Science, 69,* 275–290.

Millman, S. T., & Duncan, I. J. H. (2000b). Effect of male-to-male aggressiveness and feed-restriction during rearing on sexual behaviour and aggressiveness towards females by male domestic fowl. *Applied Animal Behaviour Science, 70,* 63–82.

Millman, S. T., & Duncan, I. J. H. (2000c). Strain differences in aggressiveness of male domestic fowl in response to a male model. *Applied Animal Behaviour Science, 66,* 217–233.

Millman, S. T., Duncan, I. J. H., & Widowski, T. M. (2000). Male broiler breeder fowl display high levels of aggression towards females. *Poultry Science, 79,* 1233–1241.

Mills, A. D., Duncan, I. J. H., Slee, G. S., & Clark, J. S. B. (1985). Heart rate and laying behaviour in two strains of domestic chicken. *Physiology and Behaviour, 35,* 145–147.

Mills, A. D., Wood-Gush, D. G. M., & Hughes, B. O. (1985). Genetic analysis of strain differences in pre-laying behaviour in battery cages. *British Poultry Science, 26,* 182–197.

Mitchell, M. A., Carlisle, A. J., Hunter, R. R., & Kettlewell, P. J. (2000). The responses of birds to transportation. *Proceedings of the 21st World's Poultry Congress* (Paper S3.13.05). Montreal, Canada: World's Poultry Science Association.

Mitchell, M. A., & Kettlewell, P. J. (1993). Catching and transport of broiler chickens. In C. J. Savory & B. O. Hughes (Eds.), *Proceedings of the Fourth European Symposium on Poultry Welfare* (pp. 219–229). Potters Bar, England: Universities Federation for Animal Welfare.

Muir, W. M., & Craig, J. V. (1998). Improving animal well-being through genetic selection. *Poultry Science, 77,* 1781–1788.

Newberry, R. C. (1992). Influence of increasing photoperiod and toe clipping on breast buttons of turkeys. *Poultry Science, 71,* 1471–1479.

Newberry, R. C., Webster, A. B., Lewis, N. J., & Van Arnam, C. (1999). Management of spent hens. *Journal of Applied Animal Welfare Science, 2,* 13–29.

Nicol, C. J. (1987a). Behavioural responses of laying hens following a period of spatial restriction. *Animal Behaviour, 35,* 1709–1719.

Nicol, C. J. (1987b). Effect of cage height and area on the behaviour of hens housed in battery cages. *British Poultry Science, 28,* 327–335.

Norgaard-Nielsen, G. (1990). Bone strength of laying hens kept in an alternative housing system, compared with hens in cages and on deep litter. *British Poultry Science, 31,* 81–89.

North, M. O., & Bell, D. D. (1990). *Commercial chicken production manual* (4th ed.). New York: Chapman & Hall.

Owings, W. J., Balloun, S. L., Marion, W. W., & Thomson, G. M. (1972). The effect of toe-clipping turkey poults on market grade, final weight and percent condemnation. *Poultry Science, 51,* 638–641.

Raj, A. B. M. (1993). Stunning procedures. In C. J. Savory & B. O. Hughes (Eds.), *Proceedings of the Fourth European Symposium on Poultry Welfare* (pp. 230–236). Potters Bar, England: Universities Federation for Animal Welfare.

Raj, A. B. M. (1998). Welfare during stunning and slaughter of poultry. *Poultry Science, 77,* 1815–1819.

Raj, A. B. M. (2000). Recent developments in stunning of poultry. *Proceedings of the 21st World's Poultry Congress* (Paper S3.13.03). Montreal, Canada: World's Poultry Science Association.

Renema, R., & Robinson, F. E. (2000). Reproductive implications of full feeding female meat-type poultry parent stocks. *Proceedings of the 21st World's Poultry Congress* (Paper S3.3.02). Montreal, Canada: World's Poultry Science Association.

Richter, F. (1954). Experiments to ascertain the causes of feather-eating in the domestic fowl. *Proceedings of the 10th World's Poultry Congress* (pp. 258–262). Edinburgh, Scotland: World's Poultry Science Association.

Savory, C. J. (1989). Stereotyped behavior as a coping strategy in restricted-fed broiler breeder stock. In J. M. Faure & A. D. Mills (Eds.), *Proceedings of the Third European Symposium on Poultry Welfare* (pp. 261–264). Tours, France: French Branch of the World Poultry Science Association.

Savory, C. J., Hocking, P. M., Mann, J. S., & Maxwell, M. H. (1996). Is broiler breeder welfare improved by using qualitative rather than quantitative food restriction to limit growth. *Animal Welfare, 5,* 105–127.

Savory, C. J., & Maros, K. (1993). Influence of degree of food restriction, age and time of day on behaviour of broiler breeder chickens. *Behavioural Processes, 29,* 179–190.

Savory, C. J., Wood-Gush, D. G. M., & Duncan, I. J. H. (1978). Feeding behaviour in a population of domestic fowl in the wild. *Applied Animal Ethology, 4,* 13–27.

Singer, P. (1990). *Animal liberation* (2nd ed.). New York: Avon Books.

Tauson, R. (1980). Cages: how could they be improved? In R. Moss (Ed.), *The laying hen and its environment* (pp. 269–299). Boston: Martinus Nijhoff.

Tauson, R. (1989). Cages for laying hens: Yesterday and today—Tomorrow? In J. M. Faure & A. D. Mills (Eds.), *Proceedings of the Third European Symposium on Poultry Welfare* (pp. 165–181). Tours, France: French Branch of the World Poultry Science Association.

Webster, A. B. (1995). Immediate and subsequent effects of a short fast on the behavior of laying hens. *Applied Animal Behaviour Science, 45,* 255–266.

Webster, A. B. (2000). Behavior of White Leghorn laying hens after withdrawal of feed. *Poultry Science, 79,* 192–200.

Webster, A. B., Fletcher, D. L., & Savage, S. I. (1996). Humane on-farm killing of spent hens. *Journal of Applied Poultry Research, 5,* 191–200.

Webster, A. J. F. (1995). *Animal welfare—A cool eye towards Eden.* Oxford, England: Blackwell.

Weeks, C., & Nicol, C. (2000). Poultry handling and transport. In T. Grandin (Ed.), *Livestock handling and transport* (pp. 363–384). Wallingford, England: CAB International.

Wood-Gush, D. G. M. (1956). The agonistic and courtship behaviour of the Brown Leghorn cock. *British Journal of Animal Behaviour, 4,* 133–142.

Wood-Gush, D. G. M. (1972). Strain differences in response to sub-optimal stimuli in the fowl. *Animal Behaviour, 20,* 72–76.

Wood-Gush, D. G. M., & Gilbert, A. B. (1975). The physiological basis of a behaviour pattern in the domestic hen. *Symposium of the Zoological Society of London, 35,* 261–276.

Woolley, S. C., & Gentle, M. J. (1988). Physiological and behavioural responses of the domestic hen to hypoxia. *Research in Veterinary Science, 45,* 377–382.

Zuidhof, M. J., Robinson, F. E., Feddes, J. J. R., Hardin, R. T., Wilson, J. L., McKay, R. I., & Newcombe, M. (1995). The effects of nutrient dilution on the well-being and performance of female broiler breeders. *Poultry Science, 74,* 441–456.

JOURNAL OF APPLIED ANIMAL WELFARE SCIENCE, 4(3), 223–234

Assessing the Welfare of Dairy Cattle

Jeffrey Rushen

Dairy and Swine Research and Development Centre
Agriculture and Agri-Food Canada

This article suggests that health problems represent some of the main threats to the welfare of dairy cattle. Although disagreement often arises about what constitutes the main welfare problems, there is broad agreement that welfare is reduced by health problems. In recent decades, a marked increase has occurred in the incidence of various production diseases in dairy cattle of which lameness is the most prevalent. This article summarizes the evidence showing that lameness is affected by the genetics of the animal and by housing. High levels of production do not necessarily lead to increased lameness, although genetic correlations between levels of production and the incidence of lameness suggest that continued high selection for milk production will likely exacerbate the problem. Denying access to pasture may increase the incidence of some forms of lameness. Both the design of the stall and the type of walking surface can have a large effect on the incidence of hoof problems for the nonhuman animal kept in free-stall housing. Finally, management and nutritional factors can have a large effect, often obscuring the influence of housing. The behavior of the cow, particularly time spent lying or standing, can influence the likelihood of lameness.

Discussions of animal welfare often stress the disagreements that exist between scientists and farmers or industry groups or between different scientists. Recently, Fraser, Weary, Pajor, and Milligan (1997) presented an excellent analysis of the varying concepts of animal welfare held by different groups and the role played by underlying values in shaping these concepts. This is a useful antidote to the all-too-convenient view that a purely objective assessment of animal welfare is possible. Although agreeing with this analysis, I suggest that, nevertheless, considerable agreement exists as to what constitutes poor welfare. This is particularly so when the focus is on welfare problems that result from health

Requests for reprints should be sent to Jeffrey Rushen, Dairy and Swine Research and Development Centre, Agriculture and Agri-Food Canada, P.O. Box 90, 2000 Road 108 East, Lennoxville, Quebec, Canada J1M 1Z3. E-mail: rushenj@em.agr.ca problems. With the research emphasis that has been placed on the behavioral aspects of good welfare, a risk lies in overlooking health problems as some of the main challenges to the welfare of dairy cattle and other animals on the farm. In this article, I focus my attention on the health problems of dairy cattle.

LAMENESS AND PRODUCTION DISEASES IN DAIRY CATTLE

Over the last decades, a marked increase has occurred in the incidence of production-related diseases of dairy cows. This is most apparent in the reported incidence of lameness in dairy herds of the United Kingdom. Surveys done before 1980 reported an incidence of less than 10% in a lactation (Russell, Rowlands, Shaw, & Weaver, 1982), whereas surveys done after 1980 reported a mean incidence of more than 20% (Clarkson et al., 1996; Kossaibati & Esslemont, 1999; Whitaker, Kelly, & Smith, 1983, 2000). Many countries now report a high incidence of lameness in dairy cattle (Barkema, Westrik, van Keulen, Schukken, & Brand, 1994; Harris et al., 1988; Philipot, Pluvinage, Cimarosti, Sulpice, & Bugnard, 1994; Wells, Trent, Marsh, McGovern, & Robinson, 1993). Guard (1999) provided an estimate for the United States of 38% per year. Lameness can be a cause of severe pain (Whay, Waterman, & Webster, 1997), and the United Kingdom's Farm Animal Welfare Council (FAWC, 1997) recently stated that lameness was the most important animal welfare problem for the dairy cow.

Lameness also is a major economic problem. Lameness reduces milk production directly by between 1.5 kg per day and 5 kg per day for 2 to 7 weeks (Esslemont, 1990; Rajala-Schultz, Gröhn, & McCulloch, 1999), and milk must be discarded when antibiotic treatment is used. The most costly effects of lameness result from its association with reproductive problems (Collick, Ward, & Dobson, 1989; Lee, Ferguson, & Galligan, 1989; Peeler, Otte, & Esslemont, 1994). Lameness is an important cause of culling, either through its direct effects or by its reducing reproduction (Rajala-Schultz & Gröhn, 1999). In Quebec, 1.87% of cows are culled explicitly because of foot and leg problems (Durr, Monardes, Cue, & Philipot, 1997), representing about 8% of all culling, although the number culled because of the indirect effects on reproduction is likely to be much higher. Over all parities, the number culled for foot and leg problems more than doubled between 1981 and 1994; for older cows of Parity 5 and 6, there was a four- to five-fold increase in the amount of culling because of foot and leg problems (Durr et al., 1997). Lameness reduces feeding time (Hassal, Ward, & Murray, 1993) and the amount of movement around the barn (Singh, Ward, Lautenbach, & Murray, 1993) as well as possibly reducing rumination time when the cows are standing (Hassal et al., 1993). Estimates of the total cost of lameness in European countries, taking into account all the direct and indirect effects, average several hundred dollars per cow (Enting, Kooij, Dijkhuizen, Huirne, & Noordhuizen-Stassen, 1997; Kossaibati & Esslemont, 1997). Recently, Guard (1999) estimated that in the United States the costs due to lameness in a herd of 100 cows represent $7,600 per year.

The incidence of other production diseases also has increased. For example, in 1980 more than 16% of cows culled were "voluntarily" culled due to low milk production (Durr et al., 1997). In 1994, however, the number of cows voluntarily culled was reduced to only 4.5%, mainly because of a large increase in premature culling for problems with reproduction (8%), mastitis (5%), and lameness (4%) or because of an unspecified illness (4.5%; Durr et al., 1997). Similar rates of culling for these production diseases have been reported from countries as diverse as the United Kingdom (Whitaker et al., 2000), the United States (Ruegg, Fabellar, & Hintz, 1998), Finland (Rajala-Schultz & Gröhn, 1999), France (Seegers, Beaudeau, Fourichon, & Bareille, 1998), and Australia (Stevenson & Lean, 1998), again suggesting a widespread trend. These production diseases are now considered the most important welfare problems facing dairy cattle (FAWC, 1997) and represent a major financial loss to the dairy cattle industry (Bennett, Christiansen, & Clifton-Hadley, 1999; Enting et al., 1997).

The causes of lameness are likely to be many. In the most recent survey, Kossaibati and Esslemont (1997) estimated that 69% of the cases of lameness in lactating cows are caused by problems with the horn of the foot, sole ulcers, abscesses and penetration of foreign bodies, white line disease, and horn overgrowth—sole bruising and heel horn erosion being some of the main causes. Infectious diseases associated with skin lesions of the hoof, such as digital dermatitis, represented 36% of cases; upper leg problems accounted for 2%.

IS INTENSIFICATION THE PROBLEM?

Critics of modern animal agriculture often point to the intensification of the dairy industry as a cause of reduced welfare of dairy cattle (Adcock & Finelli, 1995). Certainly, the incidence of lameness appears higher in the United Kingdom (Whitaker et al., 2000), the United States (Wells et al., 1993; Wells, Trent, Marsh, Williamson, & Robinson, 1995), and the Netherlands (Barkema et al., 1994)—countries with more intensive dairy production, that is, higher production per cow, larger farms, and more indoor housing—than in Switzerland (Frei, Frei, Stark, Pfeiffer, & Kihm, 1997), which has smaller farms, or Australia (Harris et al., 1988), which has lower production and less use of indoor housing. In a number of European countries, sales of "organic" milk have greatly increased in an attempt to meet the growing consumer demand for milk from less intensive agricultural systems; these organic production systems claim to pay greater attention to animal welfare issues (Sundrum, 2001). However, surveys to date show that such organic dairy farms do no better than conventional ones in terms of the incidence of these production diseases (Vaarst, Hindhede, & Enevoldsen, 1998; Weller & Bowling, 2000).

Farm surveys of the incidence of such maladies always document considerable variation between farms. For example, Whitaker et al. (2000) reported that the incidence of culling for reproductive problems, mastitis, and lameness in the highest quartile of United Kingdom dairy farms was 26% but was less than 2% in the lowest quartile. Although there is likely to be some effect because of differences between farms in the reporting of diseases, such differences also may reflect differences between farms in various factors related to the genetics and housing of the cows. In the following sections, I look at the evidence that relates the incidence of various production diseases, with a particular emphasis on lameness to levels of production per se, genetics, herd size, and type of housing.

EFFECTS OF INCREASED PRODUCTION
AND GENETIC CONTRIBUTIONS

A number of studies have investigated whether the occurrence of such maladies among dairy cows is associated with a high level of milk production. Unfortunately, such results are not easy to interpret. The occurrence of these maladies can result in lower milk production (Deluyker, Gay, Weaver, & Azari, 1991) or a correlation between high production, and a high treatment rate of diseases may reflect a higher level of vigilance and concern for disease among better herd managers (Emanuelson & Oltenacu, 1998). High levels of milk production are widely associated with high levels of mastitis (Emanuelson & Oltenacu, 1998; Faye et al., 1997; Waage, Sviland, & Ødegard, 1998). However, such a relation is not as obvious for lameness. Early studies reported higher levels of lameness in herds with higher levels of milk production (Deluyker et al., 1991; Enevoldsen & Gröhn, 1991a, 1991b), but recent studies have not reported such a relation (Vaarst et al., 1998; Whitaker et al., 2000). However, increases in production levels that result in low energy balance after calving have been associated with increased lameness (Collard, Boettcher, Dekkers, Petitclerc, & Schaeffer, 2000).

To understand the correlation between levels of milk production and lameness (or indeed of any production disease), it is necessary to separate the effects of genetics from those of management on levels of production. Rauw, Kanis, Noordhuizen-Stassen, and Grommers (1998) reviewed a number of studies that separately calculated the genetic correlations and the environmental correlations between milk production levels and the incidence of various diseases. Environmental correlations were generally absent. In contrast, there were small, but important, positive genetic correlations between the level of milk production and the incidence of ketosis, mastitis, and lameness. Studies done more recently for lameness have confirmed this finding (Van Dorp, Dekkers, Martin, & Noordhuizen,

1998). These results suggest that management differences between farms that result in differing levels of production need not necessarily result in changes in the incidence of mastitis or lameness, providing that they do not worsen energy balance after calving (Collard et al., 2000). In contrast, continued breeding and selection of animals purely based on high milk production very likely could result in an increased incidence of such production diseases. Rauw et al. (1998) suggested that strong artificial selection for a trait such as milk production leads to the animal's using its biological resources to the maximum, leaving few resources to respond to other demands or to various stressors.

The finding that many of these maladies are heritable (even if heritabilities are relatively low) has renewed interest in including such health parameters in selection indexes for dairy cattle (Boettcher, Dekkers, Warnick, & Wells, 1998). However, it remains to be seen whether the genetic correlations between high production and health problems can be uncoupled or whether they are inevitable, as Rauw et al.'s (1998) argument suggested.

That Holstein cattle suffer from a higher incidence of both mastitis (Washburn, White, Green, & Benson, 1998) and lameness (Alban, 1995; Harris et al., 1988) than Jersey cattle emphasizes the role of genetic factors in production diseases. In conventional and organic dairy herds in Denmark, a higher incidence of acute sole hemorrhages was found among Danish Holsteins than among other dual-purpose breeds (Vaarst et al., 1998). The increasing incidence of inbreeding that has been noted on U.S. farms very likely may have lead to reduced survival of dairy cattle (Thompson, Everett, & Hammerschmidt, 2000), but no evidence is available to judge whether inbreeding leads to an increased incidence of mastitis or lameness.

Concerning the effect of breeding on dairy cow welfare, the FAWC (1997) of the United Kingdom recommended the following:

> Achievement of good welfare should be of paramount importance in breeding programmes. Breeding companies should devote their efforts primarily to selection for health traits so as to reduce current levels of lameness, mastitis and infertility; selection for higher milk yield should follow only once these health issues have been addressed. (p. 66)

EFFECT OF HOUSING AND ENVIRONMENT

In the last few decades, a marked change has occurred in the way that dairy cows typically are housed. The number of cows on each farm has increased markedly, and average herd size in many countries has increased notably. However, there is not strong evidence that herd size per se increases the incidence of various production disorders. A survey of a number of farms in the United States found no evidence that culling rates for lameness, mastitis, or reproductive problems (U.S. Department of Agriculture, Animal and Plant Health Inspection Service [USDA], 1996) were higher on larger farms.

In other studies, however, large herd size has been associated with an increased risk of digital dermatitis (Wells, Garber, & Wagner, 1999), lameness (Alban, 1995), and mastitis in heifers (Waage et al., 1998). It is understandable that larger herds, leading to increased contact between larger numbers of animals, would increase the risk for infectious diseases. However, it is clear that large herd size alone should not be assumed as associated with poorer levels of welfare in general. In contrast, crowding clearly does lead to a higher incidence of production diseases. A reduced number of feeding spaces per cow have been found associated with an increased incidence of mastitis (Barkema et al., 1999), and a reduced number of cubicles per cow leads to higher incidence of hoof lesions (Leonard, O'Connell, & O'Farrell, 1996).

In some countries, particularly the United States and Canada, this increase in herd size has occurred alongside a change in the type of housing used for dairy cows. In particular, the development of free-stall housing (cubicles) has resulted in a reduction in the number of cows having access to pasture. For example, in the United States, 50% of cows now have no access to pasture, and 12% have no access at all to an outside area (USDA, 1996).

Considerable research supports the idea that housing can affect the incidence of lameness and other production diseases. Removing or limiting access to pasture increases the incidence of mastitis (Barkema et al., 1999; Waage et al., 1998; Washburn et al., 1998) and digital dermatitis (Wells et al., 1999). Although free stalls may seem preferable to tie stalls in terms of greater freedom of movement, the use of free stalls is associated with a higher incidence of lameness (Ingvartsen & Andersen, 1993; Rowlands, Russell, & Williams, 1983; Whitaker et al., 2000). Although loose housing on deep straw reduces the incidence of lameness compared to free stalls (Whitaker et al.), the reverse relation is true for mastitis (Faye et al., 1997; Whitaker et al.).

Even considering only hoof problems, the effects of housing can be complex. Compared to straw yards, free-stall housing increases sole hemorrhages but decreases heel horn erosion (Livesey, Harrington, Johnston, May, & Metcalf, 1998); housing heifers indoors compared to giving them access to dry lots increases hemorrhaging on some parts of the horn but decreases it on others (Vermunt & Greenough, 1996). That different housing systems affect different welfare problems in different ways is a common finding in research on animal welfare (Rushen & de Passillé, 1992) and makes it difficult to generalize about the effect of housing on animal welfare. The effect of type of housing on welfare can depend to a great extent on the details of the particular system. For example, cows housed in tie stalls with concrete floors had a higher incidence of sole hemorrhages than cows housed in tie stalls with rubber mats (Bergsten & Frank, 1996). Stalls that are too small or designed in a way that makes it difficult for the cow to get up and lie down also have been associated with an increased incidence of lameness or sole disorders (Faull et al., 1996; Philipot et al., 1994).

Most commentators believe that the main walking surfaces for cows play a major role in affecting the incidence of lameness, with concrete surfaces being seen as a particular problem. Philipot et al. (1994) assessed risk factors for chronic and subacute laminitis as well as heel horn erosion and found that conditions such as high steps were risk factors linked to subacute laminitis. Inappropriate flooring can increase the incidence of lameness by causing excessive and uneven wear of the hoof, by direct damage as a result of uneven surfaces or protrusions, or by causing skin breaks that increase the risk of infectious diseases such as foot rot. An epidemiological survey of farms concluded that lameness was prevalent where walking surfaces were smooth concrete (Faull et al., 1996). Cow walking surface also has been related to the incidence of apparently infectious foot diseases. Perhaps because the abrasive properties of the flooring caused physical damage to the hooves, the incidence of papillomatous digital dermatitis (foot warts) was found to be substantially higher on farms where cows walked on grooved concrete than on farms where cows walked on dirt, pasture, or smooth concrete (Wells et al., 1999).

ROLE OF BEHAVIOR

The behavior of the cattle can be an important mediating factor in influencing how housing systems lead to lameness. Colam-Ainsworth, Lunn, Thomas, and Eddy (1989) found that the incidence of lameness was high in one group of heifers who showed reduced time lying down, presumably because the stalls were uncomfortable. In more systematic research, Leonard, O'Connell, and O'Farrell (1994) and Leonard et al. (1996) found that when cows were kept either in uncomfortable stalls or with too few stalls per cow, the time spent lying down was reduced, and the incidence of hoof lesions increased. Furthermore, there were significant negative correlations across cows between lying time and the incidence of hoof lesions.

This correlation likely was due to reduced lying time leading to increased hoof problems, rather than the reverse, because lame cows in the pasture lie down longer than healthy cows (Hassal, Ward, & Murray, 1993). This correlation between reduced lying time and increased hoof problems recently has been confirmed (Chaplin, Ternent, Offer, Logue, & Knight, 2000).

Rest clearly is an important behavior for dairy cattle (Munksgaard & Lovendahl, 1993) but can be affected greatly by the design of stalls (Haley, Rushen, & de Passillé, 2000). In a survey of dairy farms in the United Kingdom, Faull et al. (1996) noted the widespread use of stalls that were either too small or poorly designed. It seems possible that by reducing resting time in cattle, the use of such stalls is a contributing factor to the high incidence of lameness.

The importance of walking surfaces in affecting lameness suggests that detailed study of the gait of cattle on different surfaces (Phillips et al., 2000; Phillips & Morris, 2000) will be particularly useful in trying both to improve methods of housing and to detect lameness.

OTHER FACTORS

Although I have discussed lameness in relation to genetics and housing, it is important to stress that nutrition and management can play a major role in influencing the incidence of lameness. For example, low energy balance following calving has been implicated in the occurrence of many production diseases, including lameness (Collard et al., 2000). The effect of these factors can interact with, or override the effect of, housing. For example, acute laminitis can occur as a result of the cow's consuming excessive quantities of grain, and poor environment can worsen the dietary causes of laminitis. Livesey et al. (1998) reported that white line hemorrhages due to increased concentrate diet were worse for cows kept on concrete than on rubber mats. Hygiene procedures such as cleaning hoof-trimming equipment between use can affect the incidence of digital dermatitis (Wells et al., 1999). The effect of such factors makes it difficult to find clear differences between housing types.

CONCLUSIONS

Production diseases such as lameness and mastitis represent the most serious welfare problems for dairy cattle. The incidence clearly is affected by a combination of genetics and housing, which perhaps is no surprise. The mounting evidence of genetic correlations between high milk production and the incidence of these maladies, combined with the worrying trend to higher levels of inbreeding in dairy cattle herds, suggests that some thought needs to be given to selecting animals primarily on the basis of health-related traits. However, the housing environment clearly is important, and that cattle genetically selected for high milk production are more susceptible to these maladies emphasizes the importance of ensuring good housing conditions for high-producing animals. Because the relation between housing and the incidence of these maladies is not a simple one, few generalizations can be made regarding types of housing such as tie stalls and cubicles. Much depends on the particular details of each type of housing, such as the number of stalls available and their size and design. Although this review focused on the dairy cow, many of the factors leading to the welfare problems discussed are typical of other high-producing animals such as pigs and poultry (Rauw et al., 1998).

ACKNOWLEDGMENTS

I thank Anne Marie de Passillé and Dan Weary for very useful discussions on this topic.

Parts of this article were presented at the Symposium on Food Animal Husbandry and the New Millennium, Philadelphia, November 1999, and at a conference on Health and Welfare of Animals on the farm, Mariensee, Germany, September 2000, Lennoxville Contribution No. 699.

REFERENCES

Adcock, M., & Finelli, M. (1995). *The dairy cow: America's "foster mother."* Retrieved August 1, 2000 from the World Wide Web: http://www.hsus.org/programs/farm/mcarthur-foster101299.html

Alban, L. (1995). Lameness in Danish dairy cows: Frequency and possible risk factors. *Preventive Veterinary Medicine, 22,* 213–225.

Barkema, H. W., Schukken, Y. H., Lam, T. J., Beiboer, M. L., Benedictus, G., & Brand, A. (1999). Management practices associated with the incidence rate of clinical mastitis. *Journal of Dairy Science, 82,* 1643–1654.

Barkema, H. W., Westrik, J. D., van Keulen, K. A. S., Schukken, Y. H., & Brand, A. (1994). The effects of lameness on reproductive performance, milk production and culling in Dutch dairy herds. *Preventive Veterinary Medicine, 20,* 249–259.

Bennett, R. M., Christiansen, K., & Clifton-Hadley, R. S. (1999). Estimating the costs associated with endemic diseases of dairy cattle. *Journal of Dairy Research, 66,* 455–459.

Bergsten, C., & Frank, B. (1996). Sole haemorrhages in tied primiparous cows as an indicator of periparturient laminitis: Effects of diet, flooring and season. *Acta Veterinaria Scandinavia, 37,* 383–394.

Boettcher, P. J., Dekkers, J. C. M., Warnick, L. D., & Wells, S. J. (1998). Genetic analysis of clinical lameness in dairy cattle. *Journal of Dairy Science, 81,* 1148–1156.

Chaplin, S. J., Ternent, H. E., Offer, J. E., Logue, D. N., & Knight, C. H. (2000). A comparison of hoof lesions and behaviour in pregnant and early lactation heifers at housing. *Veterinary Journal, 159,* 147–153.

Clarkson, M. J., Downham, D. Y., Faull, J. W., Manson, F. J., Merritt, J. B., Murray, R. D., Russell, W. B., Sutherst, J. E., & Ward, W. R. (1996). Incidence and prevalence of lameness in cattle. *The Veterinary Record, 138,* 563–567.

Colam-Ainsworth, P., Lunn, G. A., Thomas, R. C., & Eddy, R. G. (1989). Behaviour of cows in cubicles and its possible relationship with laminitis in replacement dairy heifers. *Veterinary Record, 125,* 573–575.

Collard, B. L., Boettcher, P. J., Dekkers, J. C. M., Petitclerc, D., & Schaeffer, L. R. (2000). Relationships between energy balance and health traits of dairy cattle in early lactation. *Journal of Dairy Science, 83,* 2683–2690.

Collick, D. W., Ward, W. R., & Dobson, H. (1989). Associations between types of lameness and fertility. *Veterinary Record, 125,* 103–106.

Deluyker, H. A., Gay, J. M., Weaver, L. D., & Azari, A. S. (1991). Change of milk yield with clinical diseases for a high producing dairy herd. *Journal of Dairy Science, 74,* 436–445.

Durr, J. W., Monardes, H. G., Cue, R. I., & Philpot, J. C. (1997). Culling in Quebec Holstein herds. 2. Study of phenotypic trends in reasons for disposal. *Canadian Journal of Animal Science, 77,* 601–608.

Emanuelson, U., & Oltenacu, P. A. (1998). Incidences and effects of diseases on the performance of Swedish dairy herds stratified by production. *Journal of Dairy Science, 81,* 2376–2382.

Enevoldsen, C., & Gröhn, Y. T. (1991a). Heel erosion and other interdigital disorders in dairy cows: Associations with season, cow characteristics, disease, and production. *Journal of Dairy Science, 74,* 1299–1309.

Enevoldsen, C., & Gröhn, Y. T. (1991b). Sole ulcers in dairy cattle: Associations with season, cow characteristics, disease, and production. *Journal of Dairy Science, 74,* 1284–1298.

Enting, H., Kooij, D., Dijkhuizen, A. A., Huirne, R. B. M., & Noordhuizen-Stassen, E. N. (1997). Economic losses due to clinical lameness in dairy cattle. *Livestock Production Science, 49,* 259–267.

Esslemont, R. J. (1990). The costs of lameness in dairy cows. In R. D. Murray (Ed.), *Proceedings of the Sixth International Symposium on Diseases of the Ruminant Digit* (pp. 237–251). Liverpool, England: British Cattle Veterinary Association.

Faull, W. B., Hughes, J. W., Clarkson, M. J., Downham, D. Y., Manson, F. J., Merritt, J. B., Murray, R. D., Russell, W. B., Sutherst, J. E., & Ward, W. R. (1996). Epidemiology of lameness in dairy cattle: The influence of cubicles and indoor and outdoor walking surfaces. *The Veterinary Record, 139,* 130–136.

Faye, B., Lescourret, F., Dorr, N., Tillard, E., MacDermott, B., & McDermott, J. (1997). Interrelationships between herd management practices and udder health status using canonical correspondence analysis. *Preventive Veterinary Medicine, 32,* 171–192.

Fraser, D., Weary, D. M., Pajor, E. A., & Milligan, B. N. (1997). A scientific conception of animal welfare that reflects ethical concerns. *Animal Welfare, 6,* 187–205.

Frei, C., Frei, P. P., Stark, D. C., Pfeiffer, D. U., & Kihm, U. (1997). The production system and disease incidence in a national random longitudinal study of Swiss dairy herds. *Preventive Veterinary Medicine, 32,* 1–21.

Guard, C. (1999). Control programs for digital dermatitis. In J. Kennelly (Ed.), *The tools for success in the new millennium: Vol. 11. Advances in dairy technology* (pp. 235–242). Edmonton, Canada: University of Alberta.

Haley, D. B., Rushen, J., & de Passillé, A. M. (2000). Behavioural indicators of cow comfort: Activity and resting behaviour of dairy cows in two types of housing. *Canadian Journal of Animal Science, 80,* 257–263.

Harris, D. J., Hibburt, C. D., Anderson, G. A., Younis, P. J., Fitspatrick, D. H., Dunn, A. C., Parsons, J. W., & McBeath, N. R. (1988). The incidence, cost and factors associated with foot lameness in dairy cattle in south-western Victoria. *Australian Veterinary Journal, 65,* 171–176.

Hassal, S. A., Ward, W. R., & Murray, R. D. (1993). Effects of lameness on the behaviour of cows during the summer. *Veterinary Record, 132,* 578–580.

Ingvartsen, K. L., & Andersen, H. R. (1993). Space allowance and type of housing for growing cattle. *Acta Agriculture Scandanavica: Animal Science, 43,* 65–80.

Kossaibati, M. A., & Esslemont, R. J. (1997). The costs of production diseases in dairy herds in England. *The Veterinary Journal, 154,* 41–51.

Kossaibati, M. A., & Esslemont R. J. (1999). The incidence of lameness in a group of dairy herds in England. *Proceedings of the British Society of Animal Science* (p. 221). Scarborough, England: British Society of Animal Science.

Lee, L. A., Ferguson, J. D., & Galligan, D. T. (1989). Effect of disease on days open assessed by survival analysis. *Journal of Dairy Science, 72,* 1020–1026.

Leonard, F. C., O'Connell, J., & O'Farrell, K. (1994). Effect of different housing conditions on behaviour and foot lesions in Friesian heifers. *The Veterinary Record, 134,* 490–494.

Leonard, F. C., O'Connell, J. M., & O'Farrell, K. J. (1996). Effect of overcrowding on claw health in first-calved Friesian heifers. *British Veterinary Journal, 152,* 459–472.

Livesey, C. T., Harrington, T., Johnston, A. M., May, S. A., & Metcalf, J. A. (1998). The effect of diet and housing on the development of sole haemorrhages, white line haemorrhages and heel erosions in Holstein hiefers. *Animal Science, 67,* 9–16.

Munksgaard, L., & Lovendahl, P. (1993). Effects of social and physical stressors on growth hormone levels in dairy cows. *Canadian Journal of Animal Science, 73,* 847–853.

Peeler, E. J., Otte, M. J., & Esslemont, R. J. (1994). Inter-relationships of periparturient diseases in dairy cows. *The Veterinary Record, 134,* 129–132.

Philipot, J. M., Pluvinage, P., Cimarosti, I., Sulpice, P., & Bugnard, F. (1994). Risk factors of dairy cow lameness associated with housing conditions. *Veterinary Research, 25,* 244–248.

Phillips, C. J. C., Chiy, P. C., Bucktrot, M. J., Collins, S. M., Gasson, C. J., Jenkins, A. C., & Paranhos da Costa, M. J. R. (2000). Frictional properties of cattle hooves and their conformation after trimming. *The Veterinary Record, 146,* 607–609.

Phillips, C. J. C., & Morris, I. D. (2000). The locomotion of dairy cows on concrete floors that are dry, wet, or covered with a slurry of excreta. *Journal of Dairy Science, 83,* 1767–1772.

Rajala-Schultz, P. J., & Gröhn, Y. T. (1999). Culling of dairy cows. Part II. Effects of diseases and reproductive performance on culling in Finnish Ayrshire cows. *Preventive Veterinary Medicine, 41,* 279–294.

Rajala-Schultz, P. J., Gröhn, Y. T., & McCulloch, C. E. (1999). Effects of milk fever, ketosis, and lameness on milk yield in dairy cows. *Journal of Dairy Science, 82,* 288–294.

Rauw, W. M., Kanis, E., Noordhuizen-Stassen, E. N., & Grommers, F. J. (1998). Undesirable side effects of selection for high production efficiency in animals on the farm: A review. *Livestock Production Science, 56,* 15–33.

Rowlands, G. J., Russell, A. M., & Williams, L. A. (1983). Effects of season, herd size, management system and veterinary practice on the lameness incidence in dairy cattle. *The Veterinary Record, 113,* 441–445.

Ruegg, P. L., Fabellar, A., & Hintz, R. L. (1998). Effect of the use of bovine somatotropin on culling practices in thirty-two dairy herds in Indiana, Michigan, and Ohio. *Journal of Dairy Science, 81,* 1262–1266.

Rushen, J., & de Passillé, A. M. B. (1992). The scientific assessment of the impact of housing on animal welfare: A critical review. *Canadian Journal of Animal Science, 72,* 721–743.

Russell, A. M., Rowlands, G. J., Shaw, S. R., & Weaver, A. D. (1982). Survey of lameness in British dairy cattle. *The Veterinary Record, 111,* 155–160.

Seegers, H., Beaudeau, F., Fourichon, C., & Bareille, N. (1998). Reasons for culling in French Holstein cows. *Preventive Veterinary Medicine, 36,* 257–271.

Singh, S. S., Ward, W. R., Lautenbach, K., & Murray, R. D. (1993). Behaviour of lame and normal dairy cows in cubicles and in a straw yard. *The Veterinary Record, 133,* 204–208.

Stevenson, M. A., & Lean, I. J. (1998). Descriptive epidemiological study on culling and deaths in eight dairy herds. *Australian Veterinary Journal, 76,* 482–488.

Sundrum, A. (2001). Organic livestock farming: A critical review. *Livestock Production Science, 67,* 207–215.

Thompson, J. R., Everett, R. W., & Hammerschmidt, N. L. (2000). Effects of inbreeding on production and survival in Holsteins. *Journal of Dairy Science, 83,* 1856–1864.

United Kingdom Farm Animal Welfare Council. (1997). *Report on the welfare of dairy cattle.* Surbiton, England: Ministry of Agriculture, Fisheries and Food.

U.S. Department of Agriculture, Animal and Plant Health Inspection Service. (1996). *Dairy 1996.* Washington, DC: Author.

Vaarst, M., Hindhede, J., & Enevoldsen, C. (1998). Sole disorders in conventionally managed and organic dairy herds using different housing systems. *Journal of Dairy Research, 65,* 175–186.

Van Dorp, T. E., Dekkers, J. C. M., Martin, S. W., & Noordhuizen, J. P. T. M. (1998). Genetic parameters of health disorders, and relationships with 305-day milk yield and conformation traits of registered Holstein cows. *Journal of Dairy Science, 81,* 2264–2270.

Vermunt, J. J., & Greenough, P. R. (1996). Sole haemorrhages in dairy heifers managed under different underfoot and environmental conditions. *British Veterinary Journal, 152,* 57–73.

Waage, S., Sviland, S., & Ødegard, S. A. (1998). Identification of risk factors for clinical mastitis in dairy heifers. *Journal of Dairy Science, 81,* 1275–1284.

Washburn, S. P., White, S. L., Green, J. T., & Benson, G. A. (1998). Reproduction, udder health and body condition scores among spring and fall calving dairy cows in pasture or confinement systems. *Journal of Dairy Science, 8*(Suppl. 1), 265.

Weller, R. F., & Bowling, P. J. (2000). Health status of dairy herds in organic farming. *The Veterinary Record, 146,* 80–81.

Wells, S. J., Garber, L. P., & Wagner, B. A. (1999). Papillomatous digital dermatitis and associated risk factors in US dairy herds. *Preventive Veterinary Medicine, 38,* 11–24.

Wells, S. J., Trent, A. M., Marsh, W. E., McGovern, P. G., & Robinson, R. A. (1993). Individual cow risk factors for clinical lameness in dairy cows. *Preventive Veterinary Medicine, 17,* 95.

Wells, S. J., Trent, A. M., Marsh, W. E., Williamson, N. B., & Robinson, R. A. (1995). Some risk factors associated with clinical lameness in dairy herds in Minnesota and Wisconsin. *The Veterinary Record, 136,* 537–540.

Whay, H. R., Waterman, A. E., & Webster, A. J. F. (1997). Associations between locomotion, claw lesions and nociceptive threshold in dairy heifers during the *pre-partum* period. *The Veterinary Journal, 154,* 155–161.

Whitaker, D. A., Kelly, J. M., & Smith, E. J. (1983). Incidence of lameness in dairy cows. *The Veterinary Record, 113,* 60–62.

Whitaker, D. A., Kelly, J. M., & Smith, S. (2000). Disposal and disease rates in 340 British dairy herds. *The Veterinary Record, 146,* 363–367.

JOURNAL OF APPLIED ANIMAL WELFARE SCIENCE, 4(3), 235
Copyright © 2001, Lawrence Erlbaum Associates, Inc.

BOOKS RECEIVED

Catanzaro, T. E. (2001). *Promoting the human–animal bond in veterinary practice*. Ames: Iowa State University Press.

Dunayer, J. (2001). *Language and liberation*. Derwood, MD: Ryce.

Griffin, D. R. *Animal minds: Beyond cognition to consciousness*. Chicago: University of Chicago Press.

Hancocks, D. (2001). *A different nature: The paradoxical world of zoos and their uncertain future*. Berkeley: University of California Press.

Hellebrekers, L. J. (Ed.). (2000). *Animal pain: A practice-oriented approach to an effective pain control in animals*. Utrecht, The Netherlands: Van Der Wees.

Hodges, J., & Han, K. (Eds.). (2000). *Livestock ethics and quality of life*. New York: CAB International.

Kraus, L., & Enquist, D. (Eds.). (2000). *Bioethics and the use of laboratory animals: Ethics in theory and practice*. Dubuque, IA: Benoit.

Monamy, V. (2000). *Animal experimentation: A guide to the issues*. New York: Cambridge University Press.

Paddle, R. (2000). *The last Tasmanian tiger: The history and extinction of the thylacine*. Cambridge, England: Cambridge University Press, 2000.

Reading, R. P., & Miller, B. (Eds.). (2001). *Endangered animals: A reference guide to conflicting issues*. Westport, CT: Greenwood.

Reinhardt, V., & Reinhardt, A. (2001). *Environmental enrichment for caged rhesus macaques*. Washington, DC: Animal Welfare Institute.

van Schaik, C. P., & Janson, C. H. (Eds.). (2000). *Infanticide by males and its implications*. Cambridge, England: Cambridge University Press.

CONTRIBUTOR INFORMATION

Content: *Journal of Applied Animal Welfare Science* (*JAAWS*) publishes articles and commentaries on methods of experimentation, husbandry, and care that demonstrably enhance the welfare of nonhuman animals. The scope is inclusive of all animals. For administrative purposes, manuscripts are categorized into the following four content areas: welfare issues arising in laboratory, farm, companion animal, and wildlife/zoo settings. Manuscripts of up to 8,000 words are accepted that present new empirical data or a re-evaluation of available data, conceptual or theoretical analysis, or demonstrations relating to some issue of animal welfare science. Occasional feature articles are accompanied by several invited critical commentaries on them, of up to 2,500 words each. In addition, the editors will publish free-standing commentaries, letters, announcements of meetings, news, and book reviews. Unsolicited submission of these is welcome.

Guidelines: All applicable governmental and institutional guidelines regarding the use of human and nonhuman animal subjects must be strictly adhered to. Because *JAAWS* is devoted to the enhancement of welfare, the editors will not publish studies that, in their view, involve significant pain, distress, harm, or injury to subjects. In the case of nonhuman animals, investigators should indicate the source and eventual disposition of subjects used and describe in detail housing and procedures. The editors strongly encourage discussion of the animal welfare and ethical issues raised by the implementation of their study and implied in the findings. Contributors are also encouraged to use language that acknowledges the individuality and integrity of members of other species. For example, where possible use gender-specific personal pronouns (*he* or *she*) and personal forms of the relative pronouns (*who*, not *which*), avoid terms such as *it* and *the organism*, and replace such phrases as "laboratory, farm, and zoo animals" with, for example, "animals in the laboratory."

Manuscript Submission: All manuscripts should be in English; the editors will provide help to authors whose first language is other than English. Because of the international and multidisciplinary source of manuscripts, authors are encouraged to avoid jargon. Prepare manuscripts according to the *Publication Manual of the American Psychological Association* (4th ed., 1994) obtainable from APA, Book Order Department, P.O. Box 92984, Washington, DC 20090–2984, or http://www.apa.org/books/pubman. Submit four copies of manuscripts. All text must be double-spaced. Place tables on separate pages. Include photocopies of all figures. Number all pages consecutively. An abbreviated style and formatting sheet is available from the coeditors.

Send manuscripts to either of the coeditors: Kenneth J. Shapiro, PhD, **PSYETA**, P.O. Box 1297, Washington Grove, MD 20880; or Stephen Zawistowski, PhD, ASPCA, 424 East 92nd Street, New York, NY 10128. Send books for review and other correspondence to Kenneth J. Shapiro at 301–963–4751 (telephone/fax); kshapiro@igc.org (e-mail). In a cover letter, authors should state that the findings reported in the manuscript have not been published previously and that the manuscript is not simultaneously under consideration.

After manuscripts are accepted, authors are asked to (a) submit camera-ready figures; (b) provide two electronic files of the article, one in MS Word and the other in ASCII format, on a 3½-in. disk (the content of the files must match exactly that of the printed, accepted, finalized manuscript); and (c) sign and return a copyright-transfer agreement.

Blind Review: To facilitate anonymous review, only the article title should appear on the first page of the manuscript. An attached cover page must contain the title, authorship, and an introductory footnote with professional titles and mailing addresses of the authors and any statements of credit or research support. Every effort should be made by the authors to see that the manuscript itself contains no clues to their identities. The editors will assign manuscripts to the relevant content area editor. All manuscripts will be scrutinized by at least two referees.

Permissions: Authors are responsible for all statements made in their work and for obtaining permission from copyright owners to reprint or adapt a table or figure, or to reprint a quotation of 500 words or more. Authors should write to original authors(s) and publisher to request nonexclusive world rights in all languages to use the material in the article and in future editions. Provide copies of all permissions and credit lines obtained.

Production Notes: Accepted manuscripts are copyedited and typeset into page proofs. Authors are asked to read proofs to correct errors and answer queries. Authors may order reprints of their articles only when they receive page proofs.